Small
Data,
Big
Disruptions

ɔER INGE FURSETH,
ɪent, BI Norwegian Business School

Small Data, Big Disruptions

How to Spot Signals of Change and Manage Uncertainty

MARTIN SCHWIRN

CAREER PRESS

This edition first published in 2021 by Career Press,
an imprint of Red Wheel/Weiser, LLC.

With offices at:
65 Parker Street, Suite 7
Newburyport, MA 01950
www.careerpress.com
www.redwheelweiser.com

ISBN: 978-1-63265-192-1

Library of Congress Cataloging-in-Publication Data available upon request.

Interior photos/images by Martin Schwirn
Interior design by Steve Amarillo / Urban Design LLC
Typeset in Adobe Minion and DM Sans

Printed in the United States of America
LB
10 9 8 7 6 5 4 3 2 1

To My Wife
Who Makes It All Possible

And Our Boys
Who Make It All Fun

Contents

Summaries

Figures

Timelines

Foreword

Attempts to see the future are as old as civilization itself. How can decision-makers foresee the coming changes in their industry or sector of the economy? This question has been raised many times by some of the greatest management scholars, and the answer contains the key to sustainable success.

A look at the Fortune 500 shows that success is indeed hard to sustain: By 2019 nearly nine out of every ten Fortune 500 companies that were on the list in 1955 had either disappeared, merged, or contracted. Today, digital platform companies are the most successful firms in the world. With their success, the market has seen destruction across industries and individual companies, but also creative construction which ushers commercial activities into a new era.

Towards the end of the '90s, the late Harvard Business School Professor Clayton Christensen asked in his book *The Innovator's Dilemma,* "Why is success so difficult to sustain?" After studying the issue for a number of years, he reached what he described as "the strangest conclusion": The principles of good management taught at business schools sow the seeds of any company's ultimate failure. One such principle was that companies should make evermore advanced versions of their products.

One company that followed this principle was the Digital Equipment Corporation, which produced expensive minicomputers. Digital was the most admired company in the world at the time, but it priced itself out of the market along with other computer companies in the early '90s. If these companies had learned to spot market signals, they may have been able to change course before it was too late.

When I first heard Christensen pose the question about sustaining success, I was a student of business, economics, and sociology in Oslo, Norway. During my postgraduate studies and annual research stays at UC Berkeley, I found different answers to the question, and while they pointed in the right direction, they were not specific enough. The concrete methods individuals and companies need in order to foresee change and what predictions those methods could generate were still elusive.

Some thirty years after Christensen launched his theory of disruptive innovation, I picked up the book *Machine, Platform, Crowd* by MIT Professors Andrew McAfee and Erik Brynjolfsson. Rather than posing the question of why success is so hard to sustain, the professors rephrased it by asking why so many of the smartest and most experienced people and the companies most affected by change are the last to see it coming? The answer? Because established companies are so proficient, knowledgeable, and caught up in the status quo that they are unable to see what's coming. Therefore, they miss the potential and likely evolution of new digital technologies.

At the time of this writing, the global impact of the COVID-19 pandemic is severe. Companies are struggling even more to understand market changes and to figure out how to change their strategies and business and operating models for both the near and distant future. Thankfully, Martin Schwirn has written a book to help them spot the market signals that will help them adapt.

Small Data, Big Disruptions provides, in my view, the best method to predict the coming changes in the marketplace. Companies need to understand change, to map out their potential implications for the near and distant future, and thereby make success easier to sustain.

Scanning, a vital technique which guides users from recognizing indications of change in the markets to responding to these changes appropriately and timely, is finally presented here in book form. Schwirn explains the four-step scanning process in detail, provides great examples, and shows readers how to apply the process to give companies and organizations valuable insights.

Small Data, Big Disruptions is one of the most current and important books on the topic of responding strategically to the increasing pace of changes in the marketplace. Its method will help decision-makers catch glimpses of the future— glimpses that they can then turn into competitive advantages.

—Dr. Peder Inge Furseth
Professor, BI Norwegian Business School, Oslo and Fulbright
Visiting Scholar at University of California Berkeley

Preface

I have spent three decades working on finding the crucial pieces of information that make decision-makers' lives easier. I spent the first decade of my professional career on traditional market research, including performing quantitative research as part of my theses at German and American universities as well as working with one of the biggest market-research companies on refining its test-market methodology. I spent enough time in the field to appreciate how difficult extracting meaningful insights from an existing marketplace is to understand the additional challenges that looking at future markets introduces. Since then I have helped decision-makers anticipate tomorrow's commercial environment. In dozens of scenario-planning and road-mapping projects, I have helped strategists immerse themselves in future opportunities and challenges. I have dedicated the past two decades of my professional life to horizon scanning to find the crucial changes in today's marketplace that will shape tomorrow's world. I have led and participated in hundreds of workshops to find strategically crucial developments and to make sense of them. On four continents and in more than a dozen countries, I have experienced the differing expectations that participants from organizations in different industries have and how they think about emerging trends and disruptive developments. I have seen how businesses, organizations, and government-related entities think about the future and what priorities they set when looking toward tomorrow's challenges.

During this period, I established a framework of best practices to make sense of the future in the here and now. Through the years, clients, business partners, and colleagues have asked me to summarize my experience and expertise to provide the approach and rationale underlying my methodology to help them avoid pitfalls and blind spots when making decisions under uncertainty. *Scanning* is the art and science of identifying the bits and pieces of information that enable the development of meaningful narratives about future worlds. These narratives guide strategic thinking. Scanning is a disciplined approach and well-structured process. It takes the guesswork out of contemplating what the future might bring. Scanning prepares organizations and individual decision-makers for future markets with the information that is available today.

I had the chance to plunge into scanning at a tumultuous time that dramatically challenged previously held assumptions. My introduction to scanning occurred just when the dot-com bubble started to burst. I had about a year to observe how signs of Enron's downfall were easy to miss and then very rapidly became only too obvious. I saw how the fallout of the Enron scandal then rippled through markets, taking down accounting firm Arthur Andersen in the process. Just months into my second year, I experienced how one dramatic event—the September 11 terrorist attacks—instantaneously reshaped the political climate and commercial environments. I started learning how to make sense of the future at a time when it became very clear how quickly, radically, and surprisingly business models, industries, and entire nations can change. As I came to understand scanning's strength, I also became aware that all these events were not as surprising as I had first believed them to be. The sudden stock-market plunge, the prominent bankruptcy filing, and a morning of terror did not arrive out of the blue.

There is no alternative to moving with the flow of time. Time is unstoppable. But there is a choice between being prepared or surprised. In an uncertain world, predictions of the future are inaccurate and misleading by definition; anticipating plausible developments though is a strategic option you can take. Scanning is best practice in avoiding surprises. Scanning holds the promise to move ahead strategically aware and prepare your organization to take full advantage of emerging worlds and changing dynamics.

I have to thank many people who supported my work on this book. Some got me started in my effort; others made sure that I stayed on track. Some helped me get to the finish line, and a selected few endured my entire journey. My mentor Dave Button and former colleague Robert Thomas took the time to look at manuscripts at various stages to provide me with valuable input. Bill Ralston, cofounder of Strategic Business Insights and coauthor of *The Scenario Planning Handbook: Developing Strategies in Uncertain Times*, provided not only topic-related considerations but also his experience in making a book become a reality.

Alex Soojung-Kim Pang, author of *Shorter, Rest,* and *The Distraction Addiction,* was always there when I needed help to understand the publishing world and provided guidance—and emotional support—for how to make sense of it. My good friend Thomas McKenna, author of *Muslim Rulers and Rebels,* planted the seed to get me started in this endeavor; without him no single word of this book might have found its way onto paper. And my agent Maryann Karinch not only navigated the publishing industry for me but also was the one who made sure that my words would actually reach an audience. Finally and ultimately, I am grateful for my family: my wife and partner in all things that matter and our two awesome and lively boys, who make me think and care even more about the future than ever before. They gave me the energy as well as the freedom to spend my mornings (and more) on this journey.

Bill Ralston, who knows a thing or two about uncertainty, describes scanning the following way: "Strategy decisions are like playing high-stakes blackjack, and scanning is the technique for counting cards. Strategy decisions are the biggest challenge for any organization because they bet the organization's future, success, and survival on a goal and course of action when there is so much uncertainty, so much information, so little good information, and a host of stakeholders working hard against you. Scanning is the pro's process for developing better insight than anyone else about strategy opportunities before making those bets." Well, then, let's take a look at the pros' process of choice.

CHAPTER ONE

Look Ahead

ANTICIPATING AND PREPARING
FOR THE FUTURE

———————————————— ■ ————————————————

The hands can't hit what the eyes can't see.

—MUHAMMAD ALI

Humans have always wished they could see into the future. Ancient Greeks visited the Oracle of Delphi, a high priestess reputed to know what would happen long before it actually occurred. Most likely, she was little more than a rambling charlatan operating under the influence of hallucinogens. Then came soothsayers who foretold the future by examining the entrails of sheep and carnival fortune-tellers who predicted the future by gazing into crystal balls or examining the lines on your hand. And there's the good old Ouija board. Let the little piece of wood answer all your anxious questions about the future.

Today big data promises to deliver what oracles and soothsayers and Ouija boards failed to produce. Science, we are led to believe, has developed the ultimate crystal ball. But here's the problem with this approach: big data needs, well, a ton of big data. But change starts with small data. The process of *scanning* lets you anticipate the future more reliably than all of the algorithms on which big-data analysis pins its promises.

I have always wanted to find the hidden clues that can add to the maelstrom called a marketplace. During my studies I became an expert at market research, uncovering the bits and pieces of information that can help decision-makers get a better handle on consumers and market dynamics. I studied quantitative market research and worked with one of the largest market-research companies on refining its testing approaches. Despite all of the miracles related to the quantitative approach, something was missing from the equation when looking into the future. That something is scanning, a methodology that connects the dots (the bits and pieces of small data) that will generate tomorrow's business failures and successes. I have worked with and on this methodology for more than two decades, servicing companies from virtually every industry.

"I didn't see that coming!" How often have you heard that statement about an event that, in hindsight, seems fairly obvious? Consider the evolution of media for recorded music, from unwieldy vinyl records and magnetic tapes to compact discs and digital files for streaming; the death of photographic film at the hands of digital cameras; or the fatal attack on the World Trade Center. Yes, from today's perch, the future seems uncertain, but your environment will change and in ways that you should learn to envision. Leaders may not own a crystal ball that reveals precisely what future perils and chances will visit their organizations, but they can use a powerful tool to anticipate tomorrow's threats and opportunities. And, of course, anticipation is the first step toward preparation.

Scanning, a proven four-step process for capturing and analyzing information from a company's external environment, helps decision-makers foresee coming changes in the marketplace:

1. Filter vital information from an avalanche of data.

2. Identify what matters most to you and your organization.

3. Prioritize the crucial changes that will shape tomorrow's marketplace.

4. Initiate strategies that move you from vulnerability to preparedness.

These steps help business leaders make the best possible strategic decisions. Even if you think you have developed a brilliant strategy, it will inevitably collide with reality. As Mike Tyson famously said, "Everyone has a plan until they get punched in the mouth." Scanning can take the sting out of that punch or, better yet, help you avoid it altogether.

Don't Fear Uncertainty

In less than ten years, the former telecommunications giant Nokia emerged from Finland to lead the mobile phone revolution. It rapidly grew to become one of the most recognizable and valuable brands in the world. At its height Nokia commanded more than half of the global market for mobile phones; only half a decade later, it had plunged to below 3%.[1] While it rapidly skyrocketed to the top of its industry, it crashed and burned even more quickly, culminating in the sale of its mobile phone business to Microsoft in 2013. Experts could write whole books about why and how that happened, but it all boils down to an inability to anticipate and prepare for the future.

An organization's future success depends on the ability to anticipate market and competitive dynamics, identify emerging opportunities, and foresee potential threats. Add Blockbuster, Borders, Eastman Kodak, and Motorola to the list of companies that went down in flames because they could not see tomorrow closing in on them. In hindsight, the events that led to their demise seem perfectly obvious.

How do you position yourself to anticipate and act on the dynamics, opportunities, and threats you might easily see if you could look back from the future with 20/20 hindsight? This book will answer that question and enable you to put foresight in your decision-making toolbox.

Small Data, Big Disruptions introduces and applies a methodology that enables organizations to prepare for future success. If you can see it, you can prepare for it. As Muhammad Ali once said, "Float like a butterfly, sting like a bee. The hands can't hit what the eyes can't see."

The introduction presents the four basic steps of scanning: filter, identify, prioritize, and initiate (FIPI). By taking these simple steps, any organization can avoid the disasters that befell yesterday's corporate dinosaurs and begin the journey to future strategic dominance.

The future is uncertain. But company leaders have to make decisions today. Your chosen strategy inevitably will meet reality. Better make sure that you had a chance to consider what the future might bring. Better you spent some time anticipating future plausibilities.

Your ability to anticipate future dynamics and markets is your insurance against surprises. Today's world is the result of yesterday's decisions. Tomorrow's fate starts with today's weak signals of change. If you want to identify emerging opportunities and foresee developing threats, why would you look at market charts that reflect yesterday's world?

Alas, companies don't have a future-predicting crystal ball or the Oracle of Delphi. They can fail spectacularly when they foresee a future that will not materialize. Worse, day-to-day operational needs can push aside long-term requirements; strategic thinking about the future is delayed. People usually mention Blockbuster, Borders, Eastman Kodak, Motorola, and Nokia when reminiscing about major corporate mistakes—people whose hindsight is by then 20/20. It is easy to become a retroactive expert in discerning the future that was to come. What seems so obvious when looking toward the past from the present was very difficult to anticipate when looking from the past toward the future.

The world is not becoming an easier place to navigate. My program's predecessor Kermit Patton emphasizes, "The likelihood of strategic plans' being blindsided by external developments increases every year with the increasing complexity and competition in the business environment."[2] Physicist Mark Buchanan points to a reason for such a development. He maintains, "With the global economy now far more integrated than it has ever been, chains of economic cause and effect reach across the world with disconcerting speed, exposing individuals, firms, and governments to a new kind of 'interdependence risk'—to the possibility that events quite far away can undermine the activities on which their security and prosperity depend."[3] The world relates, dots connect, and developments interact.

Anticipating tomorrow's world to prepare your organization for future strategic needs therefore is daunting. That is why many decision-makers eagerly gobble up sales forecasts and market size predictions. But everybody knows they will be wrong. These charts have a market because they pretend to reduce uncertainty. They provide a misleading picture of the future, though, because they build on past dynamics.

Embrace uncertainty instead. Learn to understand what future issues will be, where pain points emerge, what changes are ahead. Scanning offers a disciplined way to address the challenges that doing so presents. Four straightforward steps offer strategic foresight, anticipating and envisioning future possible developments (rather than predicting the future or forecasting developments in mathematical models). Strategic foresight addresses the uncertainties decision-makers face. Anticipation is the first step to responding correctly and in time when futures become the present. Strategic-management authority Paul Schoemaker maintains, "Anticipatory experiences develop [firms'] reflexes and skills for the future."[4] It's these reflexes and skills that you can train to step prepared into tomorrow's world. Presenting a method to anticipate developments and dynamics is what FIPI is about.

The idea is not new. When SRI International started working on its scanning program in the 1970s, James B. Smith and Pamela G. Kruzic commented, "As a result of [growing] uncertainties, managers are beginning to broaden their approach to planning by exploring what the future might look like before attempting to forecast what it will look like."[5] There was a time when long-term planning and explicitly accounting for uncertainties were important parts of strategic decision making. What happened? Acceleration of business dynamics leads decision-makers to believe that quick reactions can be better than proactive responses. And the growth of advanced analytics distracted decision-makers from tomorrow's challenges to today's fires.

Schoemaker underscores the importance of dealing with uncertainty in the business environment: "it affects the part of the business that managers very often don't even try to manage—the external environment—and this is where much of the potential value of the business is created or destroyed."[6] Patton refers to the interplay of such external developments in *science and technology, consumers and society,* and *commerce and competition* as the "Value Creation Maelstrom,"[7] a similar notion to Schoemaker's idea but packaged in a more vivid picture that hints at decision-makers' challenges more acutely.

Hugh Courtney, former professor at the University of Maryland's Smith School of Business, calls for a more deliberate approach to address uncertainties. "If you want to make better strategy choices under uncertainty, then you have to understand the uncertainty you are facing. Instead of burying uncertainties in meaningless base case forecasts—or avoiding rigorous analysis of uncertainties altogether—you must embrace uncertainty, explore it, slice it, dice it, get to know it."[8]

Four intuitive steps guide you toward making uncertainties tangible. Scanning will reveal the commercial forces that count in the future, the competitive dynamics that will matter tomorrow. The process I have put together and refined over the years will teach you the science and art of identifying those forces, grouping them in informative ways, and putting them in context. You will end up with a narrative, a message from the future. You will become aware of things you won't have considered. Then I will show you how to translate such knowledge into action, how to move from anticipation to preparation. Uncertainty is your friend. Everybody faces uncertainty, but the prepared ones will prosper.

Scanning is a foresighting tool that can immerse your mind in the future. Forecasting limits your view to a particular point of view of the future. Scanning opens your mind to the possibilities of tomorrow. Scanning doesn't give you the answers. Scanning allows you to consider all the right questions. Scanning enables you to develop the strategic responses needed in an uncertain world.

This book guides readers through the concept and process of scanning. You will learn different levels of considerations: how to select meaningful information from an increasingly large ocean of data, how to develop decision-relevant knowledge, and how to identify the crucial developments that will affect success or failure of your organization. You will see how a small number of events and developments today can paint a powerful picture of tomorrow. Finally, we will look at an intuitive way to focus on the parts of the picture that really count for you and your organization. You will use anticipation to prepare your organization for tomorrow.

Scanning offers a way to anticipate potential future dynamics and changes, to understand what plausible future markets could emerge, to ultimately prepare for what lies ahead. Schoemaker's premise holds true: "while we cannot fully know the future, we can better anticipate and prepare for the possibilities that we can foresee."[9]

First and Foremost: Get Started!

Scanning shows you the future. Properly designing, successfully implementing, and effectively running scanning operations require learning the ropes and building experience. Two things are of utmost importance though.

First, start the process! Most companies know they need time and effort to consider tomorrow. Many of them never find the time. There's just too much to do! Perhaps there is so much to do because you didn't prepare properly yesterday for today. Extinguishing fires can be tiresome. Preventing fires requires some thought though. Here we're looking at preventing fires. It might give you more time tomorrow.

Second, follow the steps in this book. They are straightforward. They move you from a chaotic, unstructured world to a structured understanding of tomorrow with actionable insights. There's no magic involved. Everyone can do it. All that is needed is discipline to stick to the steps of FIPI. Filtering captures the building blocks of your image of the future. Identifying what matters most provides the patterns and narratives of emerging dynamics—the shape of the future. They let you immerse yourself in a world that you need to become acquainted with. Prioritizing the crucial changes puts your company in this new world. How do you fare? What should you do? Finally, initiating strategies is the stage when the rubber meets the road. The final step gives you the competitive advantage.

The first two steps—*filtering* and *identifying*—look at the outside world of an organization. It's the world that will be. It's the world you have no influence on. It's the world you will have to deal with. The latter two steps—*prioritizing* and *initiating*—are about your organization's situation and abilities. It is here that you respond to emerging developments. It is here that you have the power to make the changes that get you to tomorrow. Both parts are crucial. Preparation without anticipation is a flailing attempt to do something. Anticipation without preparation is entertainment. Do both.

Perhaps some housecleaning here. *Scanning*, as I refer to it, is a part of a foresighting tool kit that can include scenario planning, for instance. It is often called horizon scanning although other terms are related; you might have heard them. *Horizon scanning* highlights the future orientation of the concept. *Peripheral vision* underscores the importance for decision-makers to keep the sidelines in view. The

European Foundation for the Improvement of Living and Working Conditions acknowledges confusion of terminology: "There has been much re-branding of technology watch, environmental scanning, forecasting and similar activities as foresight."[10] Don't let such terminologies distract you. I personally prefer the term *scanning* when qualitatively selecting and analyzing signals that extract future-relevant insights. Some refer to the process as *environmental scanning* because it highlights the focus on developments outside an organization. I feel this term sounds more like an analysis of the biological health of ecosystems. Perhaps that's just me. Scanning makes a historic connection too. In 1967, Francis Joseph Aguilar, a Harvard Business School scholar in strategic planning, first introduced the concept and strategic need in his book *Scanning the Business Environment*.

Tomayto, tomahto. "The point of foresight is not to generate accurate forecasts. . . . The point of foresight is to improve capabilities of foresight users to anticipate and deal with change—both exogenous change and the consequences of their own actions," as the European Foundation explains.[11] Scanning brings up the image of a radar screen though—a system that captures blips of information that then require human interpretation. I like that metaphor a lot.

The important thing is what scanning can do for you and for your organization. The future will remain uncertain, but strategic alternatives do not have to be. Schoemaker destroys illusion but provides comfort: "I am not saying that you can *control* the environment or *predict* the future. You can't. But all firms can learn how to *prepare* better for uncertainty and proactively manage the part of the business that they too often leave to fate."[12] I will use the four steps—*filter, identify, prioritize,* and *initiate*—to show you how you can understand future environments, prepare for all the uncertainties that you won't be able to eliminate, and then take the necessary steps to reduce vulnerabilities and leverage abilities.

Look for Small Data and Avoid the Haystack

"If I only knew what the future would look like?!" Well, what would you do then? Identifying new developments reliably is a necessity to get a handle on uncertainties. How would you know what can punch you if you're not even making the effort to look around. But identification is not sufficient. If you stand still, you only get punched with the expectation of getting punched. You want to move so that you're in a very good position not only to avoid the punch but also to sting like a bee.

Identification represents only the first step of making scanning a strategic chaperone within your organization. But you need to end up at actionable insights. You need to follow up with meaningful responses to changes in the marketplace. In between is a task of making any insights that show changes more robust, and you need to interpret what such change actually means.

René Rohrbeck, professor of strategy at Aarhus University, describes the process's outcome: "Corporate foresight is an ability that includes any structural or cultural element that enables the company to detect discontinuous change early, interpret the consequences for the company, and formulate effective responses to ensure the long-term survival and success of the company."[13] His description provides the why of scanning—to interpret the consequences and to formulate effective responses, to establish awareness and to create preparedness.

In the past, decision-makers struggled to get their hands on information. Today, you will struggle to find the right bits and pieces in the torrent of information. You need to look for the needles in the haystack—a haystack that gets bigger by the day. Resist attempts to analyze it all when looking to the future. Find the pieces of the puzzle that let you make sense of what's to come. Making sense of such a constant onslaught of potentially relevant developments can lead to analysis paralysis. Some decision-makers resemble the proverbial deer in the headlights; others simply abandon efforts to keep up with constant changes. No doubt, information access has moved from being limited to becoming overwhelming.

But what about big data and artificial intelligence (AI)? Can't you just lean back and let machines spit out information? This is a frequent sentiment I hear (not the leaning back, but number crunching as a solution). But true change starts with small data: the first time something happens is the start of something new. It's the small data we're looking for to find the big disruptions—*small data, big disruptions.*

When you're trying to understand fundamental shifts, big data not only stands in the way, but it can lead you into cul de sacs or completely astray. Author and consultant Christian Madsbjerg and colleague Mikkel Rasmussen argue, "By outsourcing our thinking to Big Data, our ability to make sense of the world by careful observation begins to wither."[14] Worse, the newfound data availability of big data creates temptations. Scholar Nassim Nicholas Taleb highlights the very problem of having vast amounts of data at one's disposal: "With big data, researchers have brought cherry-picking to an industrial level."[15] Or, in the words of Roger Martin, former dean of the Rotman School of Management at the University of Toronto, "Researchers have the ability to pick whatever statistics confirm their beliefs (or show good results) . . . and then ditch the rest."[16]

Futurist and author Alvin Toffler cautioned decades ago, "Our obsessive emphasis on quantified detail without context, on progressively finer and finer measurement of smaller and smaller problems, leaves us knowing more and more about less and less."[17] Context matters, overview counts. There are other considerations. Martin cautions that the "greatest weakness of the quantitative approach is that it decontextualizes human behaviour, removing an event from its real-world setting."[18] Human behavior, though, is at the very heart of all economic activities and market-relevant behavior. Humans do not act in a mathematically predictable way. In the end, human actions are at the heart of uncertainties you want to understand. "Human lives do not just unfold in a purely predictable fashion the way Mars orbits the sun. Contingency, idiosyncrasy, and choices—all of which allow for alternatives—play an indispensable role," Gary Saul Morson and Morton Schapiro at Northwestern University point out.[19]

Looking Ahead in a Nutshell

Change is constant; uncertainties are inevitable. No process can do away with these facts of life; accept them.

....

Anticipate change and embrace uncertainties. They are your ticket to the future.

....

Small data is where every big disruption once started. Filter small data from the sea of data. Identify the weak signals of change.

....

Turn small data into information. Put changes in context and interpret them. Make them relevant for you.

....

Find out what this information means for you and prepare your organization for the future. Position yourself to avoid being overrun; instead, get where the future will go first.

....

Scanning can become your strategic chaperone. The process of filtering, identifying, prioritizing, and initiating provides the steps that get you from anticipation to preparation.

....

CHAPTER TWO

Reimagine Tomorrow

UNDERSTANDING UNCERTAINTY

———————————————————■———————————————————

The future is already here—it's just not very evenly distributed.

—WILLIAM GIBSON

The future can strike you like lightning. It can also move toward you like a glacier. In either case, you can count on big surprises. On September 11, 2001, a plane struck the North Tower of the World Trade Center in New York. Then another hit the South Tower. Soon afterward both towers collapsed. It took a mere 102 minutes to change the geopolitical landscape forever. Almost a decade later, on March 11, 2011, an earthquake caused a meltdown in the Fukushima Daiichi nuclear plant, rattling the global energy world. Awareness of political, commercial, and technological landscapes could have anticipated these two events, both of which represent before/after markers for human civilization. Paying close attention to what's happening in the world can give you foresight into future possibilities, and foresight can enable you to prepare for them.

Lightning bolt or glacier, the future makes an unreliable companion. Technologies can take years to change the world; business strategies take time to yield results; implementation efforts need time to unfold. It took decades for the steam engine to shrink the physical world by permanently changing the way people and goods could move from place to place. The internet did not transform communication and access to information overnight. However, if you had employed the techniques embodied in scanning, you could have anticipated and prepared for the possibility that both technologies would radically transform society and business. Even if you imagined tragedies related to a terror attack and a nuclear accident, would you have details? Likely no. Would you have been able to pinpoint the location of the disasters? Perhaps, perhaps not. And although implications have unfolded in ways that couldn't be foretold, just the general awareness could have taken some surprise out of the events. The pain might have been the same, but the resulting responses likely would have looked very different.

Uncertainty is the future's erratic companion. Its antics end up affecting every strategic decision you make. Prediction is very difficult, especially about the future. However, if you pay attention to what's going on around you, if you become acutely aware of your surroundings, you can prepare for uncertainty. You can spend many hours a day putting out fires, or you can invest more of your time in scanning. Which works best in the long run: a fire extinguisher or fire retardant? While everyone else will be struggling to figure out how to get to the future, you will already be operating there.

Your Challenge

You have to operate under uncertainty. The future still needs to come into focus. Well, it even has to crystallize itself first. Nothing's written in stone. Developments across marketplaces, competitive dynamics, and technological advances are constantly shifting, continuously moving. Unforeseeable events seem to become the norm rather than the exception.

The future is a moving target; as soon as the appearance of clarity emerges, new elements and factors reshape the societal and commercial landscape. Nevertheless, you have to make decisions in the here and now; that's what you're paid for. These decisions will position your company for the future. If you don't know what you're doing, however, it will be a future without you.

But here's the rub: "All our knowledge is about the past, and all our decisions are about the future," as former Strategic Business Insights' colleague Bill Ralston and SRI International alumnus Ian Wilson point out.[1] Or in the words of Philip Tetlock, who studied forecasts and predictions, and journalist Dan Gardner,

"Our desire to reach into the future will always exceed our grasp."[2] Strategic decisions' crucial conundrum is that you don't know the issues you will encounter on the way to your goal. Scanning will not—should not—reduce the desire to look further into the future. Scanning will provide you with the narratives that show the stumbling blocks and point to the forks in the road that your path toward the future could have.

Your Choices

Strategic decision-makers are much more exposed to the whims and challenges of uncertainty than other managers. In many management environments, previous experience provides the foundation you need. The world might be complicated, but you know what cogs affect what wheels in what ways. Turning the cogs the right way will move the wheels as desired. In strategic environments, the world is complex. You don't know many of the cogs; you haven't explored all the wheels. And how they affect each other? Well, that remains to be seen.

Let's assume you know how fast your machine runs and how often it breaks down for how long. Any prediction about how many widgets you can build within a week of the order will be a good guess. Now, unfortunately, many strategic decisions are based on predictions too. But how do you predict what you don't know? Predicting future marketplace conditions is a common approach to decision making under uncertainty. I guess it is comforting. I know it is misleading. Predictions in complex worlds are wrong almost per definition.

Uncertainty is a multiheaded Hydra. "There is considerable uncertainty about what is likely to happen—not just in terms of precise timings and details, but also even more generally in terms of the fundamental directions of change."[3] Authors for the European Foundation describe the fundamental issue. What are you to do instead? "Connect together various driving forces, trends, and conditioning factors in order to envisage alternative futures (rather than predict the future)." Rafael Ramírez, professor at the University of Oxford's Scenarios Programme, and collaborators concur, "Rather than trying to predict the future, organizations need to strengthen their abilities to cope with uncertainty." Importantly, "Smart management benefits from a richer understanding of the present possibilities afforded from multiple views about possible futures."[4]

Kermit Patton finds, "The marketplace is a turbulent confluence of commercial, cultural, and technological forces. The most important tools for remaining afloat and thriving in the turbulence are a constant awareness of the changes going on around your organization and the ability to sense, make sense of, and adapt to these changes." Patton refers to this confluence as the "value-creation maelstrom" in which business opportunities and threats emerge.

And this is where you can get a leg up. "Scanning provides a framework with which a company can regularly and systematically marshal the pattern-recognition capabilities of a group of professionals to identify important changes in the business environment and evaluate them in the context of the company's strategy, competencies, and mission."[5] Scanning enables you to go beyond looking at the often deceivably calm market surfaces. Scanning will alert you to riptides, rip currents, and undertows. Developments beneath the obvious surface are where surprises lurk. They don't have to lurk; they don't have to be surprises.

Time will move corporations, organizations, and institutions into the future; there's no alternative. The question is whether you will be aware and prepared—and there is an option. Without anticipating what developments the future might bring, you're blindfolded moving forward. Without preparing your organization, you won't be able to leverage changes in the world. Anticipation without preparation is an academic luxury—at best. The future can be a friend that offers you gifts or an enemy that ambushes you; the difference lies in the energy that you are willing to expend in considering uncertainties.

But here is your conundrum. There is the conflict between "best management of existing issues" versus "best efforts to capture and prepare for potential developments." This conflict does not need to be, though. It is your choice between continuous crisis management and navigating the future strategically prepared. Since future developments will become existing issues as organizations and decision-makers move into the future, the time to deal with them is only a matter of when. Deal with them early; avoid later pain.

Why Scanning?

Where did scanning start? Francis Joseph Aguilar's 1967 book *Scanning the Business Environment* "deals with the acquisition of information concerning matters outside the company which can influence strategic decision-making within the company." He defines *scanning* as to "include not only purposeful *search* but also undirected *viewing*. Scanning is the activity of acquiring . . . information about events and relationships in a company's outside environment, the knowledge of which would assist top management in its task of charting the company's future course of action." Even at that time, the world became less comfortable for strategists. "Top management can no longer simply cope with conditions; it must deal with a *change* of conditions."[6]

I have been part of Scan, a program that offers scanning services, for more than two decades. The concept of creating a dedicated service emerged in the 1970s. According to SRI International's own marketing material from the 1980s, "In April 1978, SRI International began to install a scanning system to help identify early signs of change that could affect its worldwide operations, programs, and clients."[7] Shortly thereafter the research institution made the results of its efforts available to its clients. A November 1981 article in *Time* magazine, "A Dip into a Think Tank," reported on such a meeting and the overall process in detail.[8] Since then, the world has sped up, and developing understanding of changes has become ever-more urgent.

Management pundit Peter Drucker points to the benefits of looking broadly at developments in the marketplace. He states, "Increasingly, a winning strategy will require information about events and conditions *outside* the institution: noncustomers, technologies other than those currently used by the company and its present competitors, markets not currently served, and so on." He highlights the need and challenges of implementing an effective scanning process: "The development of rigorous methods for gathering and analyzing outside information will increasingly become a major challenge for businesses and for information experts."[9] It is a challenge, but it is also an opportunity.

William Gibson, author of the cyberpunk novel *Neuromancer* who coined the term *cyberspace*, famously pronounced, "The future is already here—it's just not very evenly distributed." In the end, I think, this is what scanning does:

rearrange today's pieces from the future that you're able to collect in a way so that a picture emerges. Scanning is a disciplined—replicable—approach to find the weak signals tomorrow sends, to redistribute knowledge about the future. And "Weak signals of an impending shift—*when recognized early enough*—can give you a head start to prepare to take advantage of it," as strategic-management scholar Rita McGrath reminds us.[10] FIPI is here to let you recognize such weak signals early—often very early.

The World You Are Facing

Decision-makers face two general types of problems they have to address. They require different types of fact-finding missions. It's a matter of big data versus small data—to simplify the solution process somewhat. There are complicated problems. Describing a task as rocket science says all there is to say about those problems. In principle, specialized expertise and experience get you where you want to go. The problem is understood, required engineering skills known, the outcome predictable. Mistakes will happen, but the same starting position will get you to the same result if you employ the same process. Principally, once all the factors and parameters are understood, the problem can be solved. The solution is repeatable. The decision can be learned. Big data and algorithms can provide guidance here; they might even take over the decision process. We know what we're looking for; we know where we want to be.

Complex problems, in contrast, have a vast network of relationships, inter-connections, and interactions between developments. Actors' intentions are not understood; you might not even know the involved actors. Whereas complicated systems will behave identically if the starting point and direction are identical, complex systems will behave differently—every time. Instead of probabilities of complicated systems, you now have to understand plausibilities. What can happen? Which dynamics could occur? We certainly don't know what we're looking for, but we might not even understand where we want to be in the future. This world is different. This is the world of uncertainty.

Strategic-management scholars Gökçe Sargut and Rita McGrath distinguish complicated and complex situations by underscoring one main consideration:

"Practically speaking, the main difference between complicated and complex systems is that with the former, one can usually predict outcomes by knowing the starting conditions. In a complex system, the same starting conditions can produce different outcomes, depending on the interactions of the elements in the system."[11] In complex systems unintended consequences become the norm as interactions between elements occur that nobody intended nor foresaw; the large number of elements that interact also play a crucial role in making understanding of dynamics difficult.

The difference between these problems is crucial not only in understanding the problem. The two types of situations require a completely different decision-making process. In complicated situations, decision-makers focus on developing a blueprint that the organization can follow to achieve a desired outcome. In complex situations, decision-makers have to value flexibility and create options to address a range of uncertainties. Problematically, in markets, complex situations are becoming the norm rather than the exception. Uncertainty is the crucial consideration.

Complexity moves decision making beyond the analytical capabilities of decision-makers. Predictions cannot provide any guardrails for decision making. Complex environments' past behavior is no guide for future behavior. That doesn't mean that decision-makers easily will let go of the comfort of predictions. Prakash Loungani, advisor in the research department of the International Monetary Fund, who has studied economic forecasts of recessions, concludes, "I'm a bit puzzled as to why so much attention is given to the point estimates for forecasts."[12]

But we are looking at a world that doesn't fit into formulas. You will have to employ very different skills. Morson and Schapiro compare the situation to describing Mars's behavior. "Orbit specification requires knowledge and mathematical ability but not judgment"—Newton's laws of motion will do the trick. But in complex environments, "we require judgment in situations of radical uncertainty."[13]

McGrath, who focuses on strategy development in volatile and uncertain market environments, maintains that complex environments will create at "an accelerated pace of change for which many will be unprepared." Her advice? "Today, leaders need to be sensing, as early as possible, the patterns that deserve their attention and make constant course-correcting adjustments."[14] The following chapters send you on the journey to uncover such patterns.

Avoid Surprises

Although some businesses still find themselves in fairly quiet niches of the marketplace, over the past two decades complexity and volatility have become rather the norm for most industries. The pace of change has accelerated, and the effect of change has become virtually all-encompassing with the advent of the internet and new forms of entrepreneurship generating startups at a faster rate than ever before. The increasing use of novel algorithmic applications, robotic advances, and the emerging Internet of Things will not only add complexity but also increase the unpredictability of markets, business models, and competitive dynamics. You'd better look over your shoulder to avoid being ambushed off guard.

The world is more connected, more interactive, more dynamic. It is less predictable. You will not be able to predict the future. But that doesn't mean that you cannot avoid surprises. University of Cambridge economist Helen Thompson perceives a fundamental change in the marketplace: "In the surreal world of post-2008 financial markets and monetary policy 'black swan' events shouldn't surprise us any more."[15] She refers to rare and unpredictable events that Taleb describes in his 2007 book *The Black Swan: The Impact of the Highly Improbable*. In fact, not only shouldn't black swans surprise us; we also can prepare our mind for most black swans that are waiting to hatch.

Business scholars Max Bazerman and Michael Watkins give us the term *predictable surprise*—"an event or set of events that take an individual or group by surprise, despite prior awareness of all of the information necessary to anticipate the events and their consequences."[16] Tetlock and Gardner highlight a common misinterpretation: "Many events dubbed black swans are actually gray [and] Black swans are not as wildly unpredictable as supposed."[17]

Some events always will be in a class by themselves. Extraordinary occurrences such as the terror attacks on September 11, 2001, in the United States or the Fukushima Daiichi nuclear disaster on March 11, 2011, are exceptional events. Numbers—9/11 in the United States and 3/11 in Japan—have become placeholders for societal shock and dismay. Nobody can predict them. But there were signs that such events could occur. There were signals. The World Trade Center had been attacked in a bombing in 1993. And al-Qaeda previously attacked US embassies and a US destroyer in 1998 and 2000, respectively. It shouldn't be a stretch of the imagination to make a potential connection. In 1986, a nuclear reactor in Chernobyl overheated with disastrous effects. And at the end of 2004, a tsunami in the Indian Ocean killed more than 200,000 people. Again, the vulnerability of some reactors, including the one in Fukushima, should have been a consideration. Would you be able to predict the details and outcomes? Certainly not. Should you be utterly surprised when it happens? Perhaps not either.

Most events won't arrive with such fury. That will make it easier for you to see the writing on the wall. The steam engine changed logistics and trade on such a scale that within decades old market behavior became obsolete. Understanding that asbestos, a previously widely used material, is poisonous turned industries upside down. The effects on industries such as construction or manufacturing were substantial. The internet has introduced such a wide range of capabilities that companies and individuals still work on developing novel applications and leveraging new approaches.

Similarly, cold fusion could herald the age of almost unlimited energy supply. Nanotechnology could prove to have even more severe health effects than asbestos. And quantum computing could lead to completely different approaches in computing. Identifying the point in time when such theoretical possibilities become plausibilities is challenging. But weak signals will guide you to better understand transition points.

In an ever-more tightly networked global community with increasingly rapidly paced changes, single-trend predictions will inevitably miss relationships and dynamics. These predictions almost have to be wrong per definition. Worse, Courtney points out, "The traditional process encourages managers, who are trying to generate point-forecast assumptions, to ignore whatever uncertainties they may find."[18] You won't find what you're looking for. But you will miss what you need to know. What you are looking for instead is far removed from predictions. You should make connections, find dynamics, and explore implications.

Where Change Is Emerging

You should look for developments in three broad areas (Figure 1). They will cover the uncertainties you should be aware of. They can reinforce each other. They can oppose each other. Dynamics unfold within and across each other.

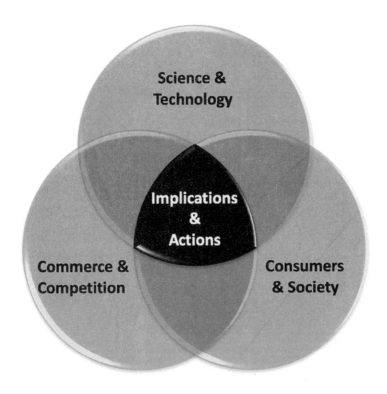

Figure 1: Areas of Relevance Highlight Dynamics

In scanning, *science and technology, consumers and society*, and *commerce and competition* establish a framework in which the whole is far more than the sum of its parts. Interactions and dynamics between and among these parts can provide a three-dimensional look at marketplaces and societies. Considering only one of the parts in isolation tends to result in misleading views of the world if not in completely inaccurate perspectives on global, societal, and commercial environments.

Science and technology captures the enabling components of commerce. This area most readily comes to mind when decision-makers try to understand change. Silicon Valley has elevated this area to a religion. Yes, it is an important source of change, so do not neglect its transformative power. After all, plastics and semiconductors have not only changed industries and consumers' lives but the entire fabric of society. But it is not the only source of change. Science and technology are enablers. Not more, not less. Innovation and change can come from all spheres of life. But as enablers they can have far-reaching effects. The authors of the *Handbook of Knowledge Society Foresight* stress, "A problem that has emerged in most science and technology-focused foresight activities is that they have only belatedly recognised the importance of broader economic, social and cultural factors."[19] Look for the big picture; consider context.

Consumers and society represents the playing field as well as the framework in which economic and commercial activities take place. This area should capture the ground in which commercial activities take place, the rules of the game. Rules take the form of regulations and laws. They also can be unwritten agreements and cultural understanding. Capturing developments here is crucial. The most promising technologically advanced materials—let's say, nanomaterials—can face market backlash. Consumers might be afraid of it. Regulators might limit their use. And they don't even need to be dangerous to encounter headwinds; perceptions here can be as strong as facts.

Commerce and competition encompasses business models, commercial conduct, and industry dynamics—areas that lie at the very heart of day-to-day activities in organizations. Novel business models alter market behavior; startups introduce competitive changes. Decision-makers usually understand their own industry quite well. Unfortunately, such familiarity can make you blind for changes in a world that you won't consider to be your backyard.

The Myth of Technology's Role in Change

The areas capture the space you want to keep tabs on. But some comments here are in order. Many misconceptions exist. High tech tends to be at the center of many scouting efforts; low tech is neglected. Long-term effects are underestimated. Industry outsiders tend to be ignored.

Let's look at *science and technology*. This area has become the representation of innovation. In some cases it can be; in most cases it's merely the source of enablers. Then you might think of developments in this area as objectively verifiable and measurable. But technologies truly become valuable only in the context of commercial use and market applications. Here subjective judgments matter though. Objectively better does not necessarily translate into commercially successful. Perception can create friction areas between objective assessment of developments in *science and technology* and subjective valuation by *consumers and society*. Or, the best technology doesn't matter if the market won't care.

Genuine technological progress might not find use if industry participants' core competence hinges on a different discipline. For instance, such a conflict existed in the photographic industry between the emerging digital technology and incumbents' chemical expertise. Eastman Kodak got caught in the conflict between the areas of *science and technology* and *commerce and competition*.

Innovation can come as readily from novel competitive approaches and consumers' changes in lifestyle as it can come from technical advances. In fact, many of the developments and changes that emerged from the Silicon Valley—a synonym for high tech—are, on closer inspection, innovations in business practices. Novel payment approaches enable new approaches to transactions, peer-to-peer technologies change distribution patterns, and online retail and auctioning sites enable massive product selections. It's really not about technology. All of these advances leverage technologies, but they do not necessarily require advanced or novel technologies.

Silicon Valley, roughly the southern part of the peninsula that lies between San Francisco and San Jose in California, is known for its technological contribution but usually underestimated for its development of new business models and commercial approaches. In truth, both developments from the start resembled a powerful marriage. When scientists Robert Noyce and Gordon Moore (who gave birth to the by-now-inescapable Moore's law) left Fairchild Semiconductor in the late 1960s to develop a new type of semiconductor, Arthur Rock played as important a role as the two founders in what was to become Intel, today's semiconductor powerhouse. (Rock had also played a role in the founding of Fairchild Semiconductor itself.) Rock was the venture capitalist who got the endeavor off to a promising start. Venture capital, providing risk money to promising startups, was a very new concept. In the following decades, the venture-capital industry learned and grew alongside—in fact, in tight connection to—the information-technology industry. The argument can be made that without venture capital, the information-technology industry, including software and web-based companies, would have gone a very different direction (and at a much slower pace).

Apple's iPhone is hailed as a major milestone in technology, but perhaps the real advance was not a technological one. Other smartphones already had some success. In 2002 Palm had already introduced a line of smartphones, the Treo series, which included similar features as the iPhone had in 2007. And many Apple enthusiasts will point to the virtual keyboard as a difference that trounced Palm. I believe that the real difference lies somewhere else. Apple had previously introduced a marketplace for music files, iTunes, which facilitated distribution of music to iPods, the company's music players. The company used a similar concept for the iPhone to establish a marketplace for vetted third-party apps that consumers were able to easily download and install. Only the wide range of apps made full use of the capabilities smartphones had in comparison to feature phones. (Palm's Treo series enabled use of apps, but finding them on the internet and installing them were a hassle.) Apple's success in the phone market is based on the development of an infrastructure for developers of apps—a new logistics concept that allowed seamless distribution of software programs to smartphones. Apple didn't introduce the smartphone to the world. Apple introduced a crucial service platform to the market. Apple promoted a business-model shift.

Technology has been in the limelight as the ultimate driver of innovation for decades. Such a notion is understandable given that technology has become an essential element in almost every application in professional and personal life. The notion of technology as the ultimate driver of innovation is therefore not wrong per se, but it is not quite right either. Technology is an enabling element of applications and functionalities. Technological advances tend to focus the attention of decision-makers and strategists because of the fascination they can generate. And a focus on technology limits the breadth of developments decision-makers might consider. Such a focus unduly reduces the world of external complexities. No doubt, *science and technology* is an important source of external developments, but only *consumers and society* and *commerce and competition* turn technological enablers into applications, technological advances into relevant business cases.

In October 1985, Dr. Melvin Kranzberg, Callaway Professor of the History of Technology at the Georgia Institute of Technology, gave an address as president of the Society for the History of Technology. "Every technical innovation seems to require additional technical advances in order to make it fully effective." And "Although technology might be a prime element in many public issues, nontechnical factors take precedence in technology-policy decisions." He points to a misconception I frequently encounter, namely the notion among strategists and technologists that objectively "better" technologies will become successful. Or, in Kranzberg's words, "Technologically 'sweet' solutions do not always triumph over political and social forces."[20]

Companies have made digitalization a key focus in recent years. Gerald Kane, professor of information systems at Boston College, offers related insight. "The true key problem facing organizations with respect to digital disruption is people—specifically, the different rates at which people, organizations, and policy respond to technological advances."[21] A holistic view is crucial. Technology is embedded in commercial activities; it does not exist in a vacuum.

Coincidentally, the German Fraunhofer Society for the Advancement of Applied Research has inaugurated a new group. The Fraunhofer Group for Innovation Research will enter "dialogue with industry, politics and society." The group acknowledges, "Structural changes through technological developments must . . . be recognized and understood at an early stage in order to actively shape the long-term economic, social, political and cultural impact on society and the economy." The society also points to the reciprocal nature of technology and other areas of the marketplace; professor Wilhelm Bauer, executive director of Fraunhofer IAO, explains, "By founding the new group, the Fraunhofer-Gesellschaft wants to further consolidate its role in researching and supporting innovation processes and the technological, economic and social conditions that shape them."[22] Technology embeds and enables. Technology does not provide meaningful benefits without an environment capable and willing to make use of it.

Know How to Expect Surprises

"Nearly all surprises have visible antecedents. However, people have a powerful tendency to ignore warning signals that contradict their preconceptions." Day and Schoemaker call for vigilance to create crucial awareness of developments that are still difficult to sense. "Leaders need vigilance—that is, a heightened state of awareness, characterized by curiosity, alertness and a willingness to act on incomplete information. . . . But leaders also need a broader type of vigilance: looking for weak and unexpected signals."[23] Finding such signals starts with getting acquainted with uncertainties. Get uncertainty on your side. Make surprises your expectations. Search for the signals that let you reimagine tomorrow.

Reimagining Tomorrow in a Nutshell

You have to operate under uncertainty. Learn to look for the weak signals that can tell you about the uncertainties you will face.

··

Complex worlds proliferate as connectivity increases and the pace of change accelerates. Use big data to help you in complicated problem situations. Trust small data to find the big disruptions you will encounter moving into the future.

··

Don't attempt to predict the future. Learn instead to expect surprises. Embrace uncertainties to anticipate tomorrow.

··

Tomorrow results from today's dynamics. See today's world in a different way to reimagine tomorrow.

··

Look for events and developments in *science and technology, consumers and society,* and *commerce and competition* to gain a holistic view of how these areas could interact in the future.

··

Then follow scanning's four simple steps to guide you toward anticipating tomorrow.

··

CHAPTER THREE

Set Expectations

KNOWING THE DIFFERENCE BETWEEN ANTICIPATION AND PREDICTION

———————————————■———————————————

Prediction is very difficult, especially about the future.

—Niels Bohr

Just as it took countless brushstrokes to create the Mona Lisa, it takes many developments to shape the future. The internet existed in its basic form, the Advanced Research Projects Agency Network (ARPANET), in the mid-1970s, but it was largely the domain of scientists and university professors. Most consumers and business leaders knew little or nothing about it. When the World Wide Web and a user-friendly web browser made the infrastructure accessible in the 1990s, it spread throughout human culture like a wildfire. Although few business executives could have predicted how the internet would change everything, some foresighted individuals did anticipate its arrival and the opportunities it might present. Some waited for the curve to arrive; others got ahead of it.

The business environment, like life itself, is chaotic, complex, and hard to control. But we do try to make it behave. To make sense out of all the brushstrokes, we look for trends. Spotting trends comforts us because they seduce us into thinking we can see what will happen tomorrow and the next day. But trends are fickle animals. All too often they bite the hand that feeds them. Suddenly, a seemingly well-behaved trend reacts erratically. A beloved strategy based on trends that marked the past has led more than a few companies astray.

The future defies simple extrapolation of current developments. First brushstrokes of a picture can be misleading. What looked like a landscape at first eventually turns into a portrait of a smiling young woman. First impressions are almost always misleading. Extrapolations of initial glimpses likely lead you in the thicket. High-profile examples of household company names led astray abound.

You cannot predict the future, but you can anticipate it. You must understand the difference and make decisions accordingly. Scanning uses brushstrokes, not to create a perfect picture of tomorrow, but to develop insights into the way circumstance, technologies, consumers, and businesses might align to create markets. Scanning delivers the sketch that evolves into the Mona Lisa.

Anticipate, Don't Predict

In 1968, Paul R. Ehrlich, a biologist at Stanford University, published *The Population Bomb: Population Control or Race to Oblivion*. Ehrlich advanced the idea that the ever-growing world population would soon run out of food. He argued vehemently for population control, going so far as to suggest adding sterilizing agents to the water supply. The book climbed the bestseller lists. The author's ideas seemed so intuitive, his proposals convincing. Well, the future didn't turn out exactly the way Ehrlich predicted. While the world's population more than doubled from some 3.5 billion people in 1968 to more than 7.5 billion people today, famines and hunger have actually declined, even though some regions still do face dire shortages.

In hindsight, this forecast seems like an obvious error. Ehrlich based his predictions on a single trend line—population growth—and existing food production methods. He failed to see some other brushstrokes, especially the human capacity to apply ingenuity to a problem. Yes, the population increased, but so did incentives to devise ways to combat hunger—and to make money in the process. The Green Revolution, sophisticated agricultural technologies, and vastly improved processing and logistics capabilities conspired to improve food production and distribution.

In other words, Ehrlich's focus on a single development in *consumers and society* prevented him from considering potential developments in *science and technology* as well as *commerce and competition*. Physician and statistician Hans Rosling puts it nicely: "Don't assume straight lines. Many trends do not follow straight lines but are S-bends, slides, humps, or doubling lines."[1]

Marina Gorbis, executive director of the Institute for the Future, delineates the difference between predicting and understanding the future: "I firmly believe that no one can predict the precise shape the future will take or which specific organizations or individuals will be successful. However, I also believe that we can see important directions of change that are driven by a confluence of larger trends and that it is important for us as individuals and as a society to understand these shifts and explore how they may reshape our lives, our organizations, and our routines."[2]

You cannot predict the future, but you can anticipate it by gathering and evaluating the right set of brushstrokes. Scanning uses brushstrokes, not to create a perfect picture of tomorrow (a prediction), but to develop insights into how developments create tomorrow's markets. It delivers (anticipates) a more complete sketch that reveals the possibility of a Mona Lisa.

Establish Context

Scanning is not a statistical game that presents tomorrow as a trend line. Scanning puts developments in context. Understanding how technologies, consumers, and businesses need to align to create markets is at the core of the methodology. Correlations, extrapolations, and predictions prove self-limiting; scanning meanwhile explores the full range of options and opportunities. Scanning focuses the view but widens the perspective.

A single data point by itself does not make for a market development. Supply-chain partners have to see their own benefits, marketplaces have to be commercially ready, and consumers have to be open for adoption. The internet existed in a basic form, the ARPANET, in the mid-1970s—without commercial impact. Only the emergence of the World Wide Web and Marc Andreessen's development of the first easily employable web browser, Mosaic, put some crucial technological pieces in place in the 1990s. Consumers not only needed a way to easily access the network but also then required time to become familiar with and confident in the use of computing platforms to enable adoption. The future arrives slowly—and then can suddenly change the world.

Business strategists also took time to understand the transformative power of the internet. There was a reason why it took the dot-com boom a full quarter century to get in swing after the first message was sent on such a network. Mosaic pieces need to fall into place; more than a couple brushstrokes are needed to complete a painting. A complete picture is needed.

Scanning does not attempt to predict the future. The reason is simple. Any attempt to do so is futile. Different concerns matter. Context is crucial.

Understand the Layers of the Onion

Every company's world resembles an onion with various layers. Every decision-maker has to work on peeling these layers. Unfortunately, while some layers will feel very familiar—perhaps too familiar to develop an objective view—other layers represent the world of uncertainty. At the core of the onion is the company itself. This is the world managers know; this is the world that's at their command. The next circle reflects the industry or market and region a company operates in. This layer is the environment most decision-makers will feel comfortable with. It's the world they usually grew up in, the world they deal with on a daily basis. Dynamics are known; competitors are old pals. The outer circle is where scanning's power comes into play. It is the layer that features activities outside of the industry or market a company is active in. Here decision-makers usually lack understanding of basics and dynamics. Surprise lurks here; disruption happens in this layer. This layer is where scanning reduces uncertainty. Scanning becomes the light that shines a spotlight on the challenges that occur in the dark.

It is here that the Blockbusters, Kodaks, and Nokias were surprised when companies in the outer layer took a stab at the heart of these companies' businesses. It is here where recording companies, mapmakers, and encyclopedia publishers didn't see what they had coming with the emergence of the internet. It is easy to look at these companies today and wonder how they failed to notice what would kill them. What's known today was yesterday's uncertainty though. It is therefore so much more surprising when a company manages the outer layer to its fullest advantage. Instead of being a deer in the headlights as the truck of the future approaches, Amazon.com understands layers and not only prepares for defending its market but leverages the changes in the outer layer (Figure 2).

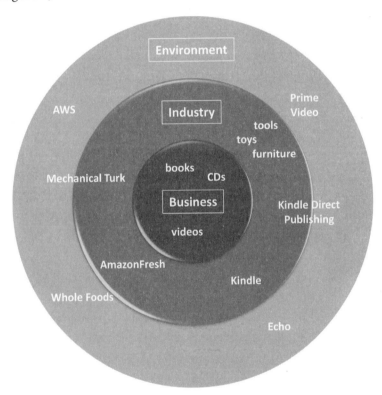

Figure 2: Amazon.com's Onion

We all know where it started. Amazon used the internet to sell mainly books, CDs, videos—packaged media—and some other product categories. Very early on, the company expanded within the retail industry to include a much wider range of products such as toys, tools, even furniture. In the end, Amazon leveraged the internet to leapfrog traditional mail-order catalog companies. In the US, one-time mail-catalog giant Sears now finds itself in permanent reorganization. Europe's biggest mail-order retailer, Quelle, does not exist anymore in its initial form. Consider that Sears and Quelle already had a massive selection, warehousing system, and logistics operation in place before Amazon even entered the scene. Wasn't it just about putting a catalog online? Someone clearly missed the boat.

But Amazon went far beyond its core layer. Amazon Prime offers streaming services among other offers. The company introduced self-publishing services, moving beyond the mere sales of books into book production. In electronics, Amazon introduced the ebook reader Kindle to drive ebook sales, but has since moved far beyond its industry layer with the introduction of its smart speaker Echo. The company expanded within online retail to food deliveries and then entered brick-and-mortar stores through the acquisition of Whole Foods and its experimentation with Amazon Go, retail outlets that test advanced technologies such as automated checkouts and frictionless payment options. Amazon even leveraged internal capabilities to offer an entire line of services completely unrelated to retail. Crowdsourcing service Mechanical Turk allows users to tap into a wide network of contract workers. And Amazon Web Services (AWS) repackaged the company's computing capabilities as an outside service for on-demand cloud computing tasks. Amazon might be an extreme example, but keeping an alert eye on developments in the outer layer can make a company successful. How successful? Amazon generates more than half of its total operating earnings from AWS,[3] the service most removed from its initial offer, far in the outer layer of Amazon's onion.

Companies need to realize that their market sphere has expanded; focusing only on familiar territory has become a proposition for failure. Scanning reveals changes in the outer layer, the layer organizations have no control over—at most very limited and indirect influence on—but can leverage to their advantage. Amazon didn't develop cloud computing but was quick to seize the moment. Identifying such a development, though, is like picking up a strand from the fabric of the future; it is worthwhile following and considering how the development can ripple through markets and industries. If all the right ingredients come together, commercial waves can become breakers that roll over long-established industries. In the wake of such tsunamis, the unprepared will find themselves flotsam and jetsam while others sail to promising shores.

Notice Changes Early

In the early 1990s, the internet had been around for almost a quarter of a century, but commercial success took time to arrive. In fact, most consumers likely didn't even know by the beginning of the 1990s that there was such a communication network. By the end of that decade e-s (as in ecommerce and ebooks) and dot-coms had changed the business world. Many industries were surprised; the music business was in shock.

On October 29, 1969, the internet's predecessor ARPANET started first operations. The University of California, Los Angeles, sent a message to the Stanford Research Institute (now SRI International) in Menlo Park, California. Its potential use for media and entertainment applications eluded all but the most visionary minds. Brushstrokes were missing; mosaic pieces still had to fall into place.

During the 1980s, project work at the European Organization for Nuclear Research, CERN, in Switzerland, led Tim Berners-Lee to introduce the World Wide Web—the *www* in internet addresses—to the world in December 1990. Meanwhile, some researchers at the Fraunhofer Institute in Germany worked on a compression method to create a music file—MP3—finalizing it in 1992. In January 1993, now venture capitalist Marc Andreessen and a colleague developed Mosaic, the first usable web browser.

A picture—a signal of change—emerged (Timeline 1). The right set of ingredients came together within not even three years—after more than two decades of waiting. The future arrives slowly—and then suddenly knocks at the door. In the following years, universities provided network access, and students got familiar with computers and the World Wide Web. Now first consumers knew about the network's capabilities. A full pattern of developments emerged; a signal of change came into focus. The entertainment world—and much more—was set up for a full transformation by the mid 1990s.

By April 1998, *Electronic Delivery and Distribution of Music* identified a service opportunity that hit dorm rooms of universities only a year later.[4] That's when—in June 1999—first beloved, now infamous Napster introduced its controversial peer-to-peer file sharing service for MP3 music files. Napster caught the music industry completely off guard. When music group Metallica sued Napster in April 2000 for copyright infringement, Napster became notorious.

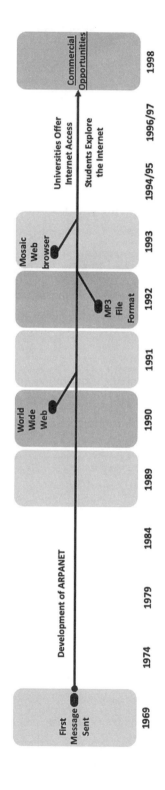

Timeline 1: Signal of Change for Music–File Distribution

As we know now, Napster went off into oblivion, but the idea of music distribution had changed the industry for good. Apple's iTunes successfully introduced legal downloads of digital music files in April 2003, and by today streaming has changed the industry's landscape again—substantially. The industry transformed within not even half a decade from the alert provided in the report.

Change was rapid. In January 2000, music-streaming service Pandora had already started its online music service. Spotify followed in April 2006. Ten years later, market-research company Ipsos found in its report on the music industry in 2016: "YouTube is the most used music service: 82% of all YouTube visitors use it for music."[5] Putting developments in a timeline highlights scanning's power in raising awareness of future possibilities (Timeline 2).

Electronic Delivery and Distribution of Music foresaw more than file sharing; it alerted to emerging revenue hiccups. It pointed to pay-to-listen, rental, subscription, and advertising models—models that the industry is still struggling with to find revenue streams that meet the expectations of all involved parties. Music-industry participants perceived the following decade very differently. Record stores for the most part went out of business, but artists, music publishers, and recording companies all faced new realities—and encountered a very different revenue landscape. "Everyone has a plan until they get punched in the mouth." By now streaming of files has made music a commodity. From luxury to commodity within a decade and a half.

As is often the case, adjacent industries go through similar pains shortly thereafter; this is where the layers of the onion gain importance. Many bystanders' smirks have become tears of regret. The movie industry is facing similar opportunities and threats with a similar move to files and streaming applications. Again, rental stores—most prominently Blockbuster and Hollywood Video—got wiped from the retail landscape. And 2018 was the first year in which the revenue that online movie offerings generated passed global box-office revenue.[6] Amazon's and Netflix's streaming services left their mark on the industry—no surprise, given the music industry's pain in the previous decade. Don't say you didn't see it coming.

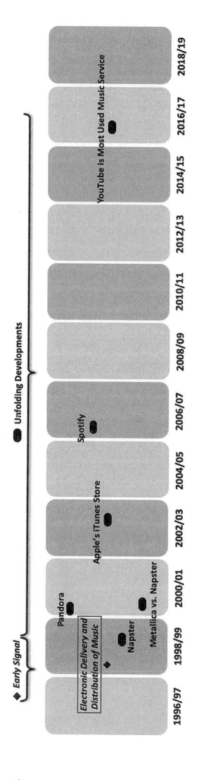

Timeline 2: Music–File Distribution

Become Aware

The best scanning process will not be able to predict the future. The interest in crystal-ball vision is understandable, but the future will remain uncertain. Although scanning does not provide the future, it certainly prepares for the future. In 1998, the lay of the land for the music business and beyond came into focus, but many decision-makers didn't widen their perspective.

Predictions and forecasts provide an unambiguous look at the future—the most dangerous way to think of the future. Decision-makers though should stay vigilant. Niels Bohr astutely remarked, "Prediction is very difficult, especially about the future." The quote identifies not only the challenge but also the limitations of statistical approaches; the earlier is understood, the latter is frequently ignored. Forecasts are alluring but make decision-makers single-minded. Sargut and McGrath sensibly advise, "In an unpredictable world, sometimes the best investments are those that minimize the importance of predictions."[7] Awareness is the strategic asset; predictions quickly turn into liabilities.

Scanning provides such awareness. Wharton School professor George Day and Schoemaker introduced the term *peripheral vision* to describe such awareness.[8] I like the term because it highlights what the process can achieve. Information to create such awareness is out there. It's just not right in front of you; it's at the periphery. Organizations looking ahead tend to fail to notice these bits of information. Business scholars Max Bazerman and Michael Watkins stress, "While individuals may recognize key pieces of the puzzle, failures of environmental scanning and information integration may prevent the organizations they belong to from perceiving dire emerging threats."[9] Scanning adds brushstrokes. Some of them will tell a story.

If it's not predictions, what then does scanning have to offer? A straightforward four-step process to identify pretrend topics and friction areas.

- *Pretrend topics*—issues that are just emerging. They are not trends yet and may never become trends. They are issues with the potential to transform today's world into tomorrow's landscape. They are early signals of change.

- *Friction areas*—areas where new developments clash with traditional market behavior, where emerging technologies challenge existing solutions, and where novel competitive dynamics question historic corporate conduct. They are market dynamics that fight each other.

Pretrend topics and friction areas together create uncertainty in the marketplace, and both create the fabric of the future. Scanning identifies them. Scanning provides a sounding board to bring them to decision-makers' attention. They enable decision-makers to focus on emerging market opportunities and developing corporate vulnerabilities. Scanning does not predict the future; scanning explicitly addresses uncertainty by highlighting the areas of uncertainty. Don't tune out uncertainties; zero in on them.

Stay Alert

There is too much information. Unprecedented access to data has led to new analytics. Big data is the natural response to overwhelming data availability. But big data quickly runs into a wall. Big data tackles complicated situations; human sensemaking can elucidate complex environments. Operations are complicated, but the world is complex.

Big data is focused on, well, big data—many data points, a lot of data points. Strategic considerations though should be focused on small data. The first email sent, the first creation of graphene, the first robot connecting to the internet, and the first time hydraulic fracturing (fracking) was used. That is the information that allows users to perceive the future of communication, new materials, smart manufacturing, and a fundamentally changed energy landscape.

Taleb highlights the dilemma: "I am not saying here that there is no information in big data. There is plenty of information. The problem—the central issue—is that the needle comes in an increasingly larger stack."[10] Smaller haystacks make needles more apparent. Focus on small sets of relevant developments. Scanning, with its focus on small sets of data, makes the haystack manageable. Martin emphasizes the "world is not responding to our attempts to control it with quantitative models. Our chaotic environment demands a new approach that pays attention to qualities in addition to quantities."[11] Exactly!

Courtney calls for a more deliberate approach to address uncertainties. "Instead of burying uncertainties in meaningless base case forecasts . . . you must embrace uncertainty, explore it, slice it, dice it, get to know it."[12] And there is a process to find the uncertainties, to embrace them. There's a process to turn uncertainties into stories and narratives to make sense of tomorrow. Scanning starts with events and developments—new technologies, market changes, novel business models—that hold the key to the future.

Leverage the Power of Scanning

Two-thirds of corporate strategists admit that they are surprised by competitive events with high impact.[13] And four out of five see a greater need for peripheral vision than what they currently have. The vast majority admit that their companies do not have a system in place to provide early warning of impending events. Why not? Scanning is an easy way to gain peripheral vision.

Scanning is a four-step process (Figure 3). It's not magic; it's discipline. Everybody can do it. And everybody should do it. Scanning includes *filtering data points for manageability, identifying patterns and developing narratives, prioritizing identified issues by considering implications*, and ultimately *initiating follow-up steps to take proactive action*. FIPI! You just took the journey from a chaotic world with innumerable uncertainties to an understanding of the developing commercial landscape that will guide you into the future.

Each step will help you further your understanding of the future. Only all of the steps will enable you to plot your pathway to the future:

- *Filter data points for manageability.* The need to focus on a selected number of events is the strength of the process, not its weakness. It directs attention to the early signals of change. It separates information from data. It allows human sensemaking. Filtering magnifies crucial information. Filtering makes sensemaking manageable. William Osler, the father of modern medicine, noted, "The value of experience is not in seeing much, but in seeing wisely." The first step, filtering, helps you see wisely.

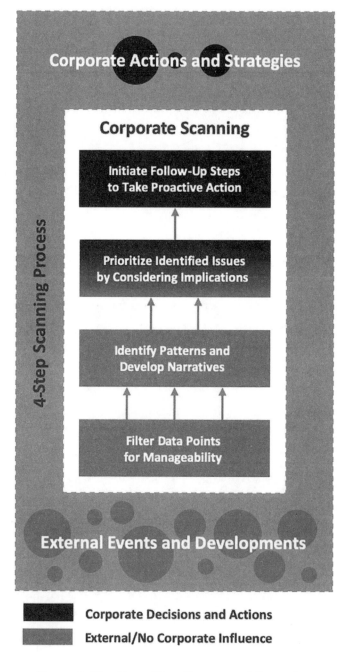

Figure 3: The Four-Step Process of Scanning

- *Identify patterns and develop narratives.* Seeing similarities, discerning conflicts, and developing a story around such dynamics are at the core of scanning; they are at the core of human sensemaking. Pattern recognition helps guide thinking and conversations. Narratives extract knowledge from patterns. Narratives can be communicated; they can be discussed. Narratives leverage and exploit human imagination. And, as Albert Einstein found, "Imagination is more important than knowledge. For knowledge is limited."

- *Prioritize identified issues by considering implications.* Not everything is relevant for everybody. We now need decision-making focus. Your own corporation, your own decisions become the center of attention. Considering implications involves looking at angles and perspectives of a potential development similarly to the way one would turn a Rubik's Cube to find one's bearing. Narratives offer anchor points for decision-makers to envision changes and transformations. Implications for your corporation readily emerge. The most important ones quickly float to the top.

- *Initiate follow-up steps to take proactive action.* Up until here, scanning created awareness of emerging issues. But decision-makers have to prepare their organizations. Now that we know what the future could look like, now that we understand the potential implications, we have to make the decisions that move the organization in a favorable place. This step is as crucial to the process as it is vulnerable to neglect. Care is necessary to take the steps needed to respond to marketplace changes.

Scanning guides you seamlessly from gathering information from the external environment (gray) to your own decisions (black). FIPI establishes awareness about developments outside of your sphere of influence to then allow you to take the steps within your control to prepare for the approaching future.

The focus on human input and analysis is by design. Intuition, imagination, and emotions are elements that big data applications miss. Humans create connections and dynamics. People and their organizations are not rational in a mathematical sense. A lump of lead and a feather fall at identical speed in a vacuum, but in the real world they behave very differently with wind movement, rising air, and turbulences affecting the fall of the feather substantially. Complex situations require human understanding.

Use the Awareness Gap to Your Advantage

When you're looking toward the future, accuracy is not—cannot be—the measurement of success. The future is uncertain after all. The future from today's perspective is about possibilities. Identifying potential developments that then won't take place is par for the course. What is the loss? You developed a better understanding of underlying dynamics in the marketplace. Missing developments that then transform the world is disastrous though. Discipline and effort in collecting a wide set of information are crucial. Get used to thinking about issues that won't happen; you will still learn a lot about the world. No effort is lost. Forecaster Paul Saffo brushes aside accuracy concerns in his tellingly titled article "Six Rules for ~~Accurate~~ *Effective* Forecasting." The goal of forward-looking activities "is not to predict the future but to tell you what you need to know to take meaningful action in the present"—"the primary goal . . . is to identify the full range of possibilities, not a limited set of illusory certainties."[14] It's about widening the perspective.

What can you get out of scanning? The short answer is: a lot! Scanning lets you exploit the *awareness gap*: the period between the moment when you could know about a new development and the moment when most of the industry— if not everybody—knows about it. It's a matter of competitive advantages—the difference between when you're ahead of the curve and when you will struggle, and likely fail, to catch up.

Figure 4 shows the gap and highlights the competitive advantage: the points in time when potential awareness is possible, when industry awareness sets in, and when awareness turns into public knowledge. In between, any awareness lead shrinks; over the course of the timeline, competitive advantages turn to strategic liabilities.

Such an awareness lead has two stages. If you do a good job at scanning, you will gain insights that are truly unique. You will be ahead of the curve, on top of

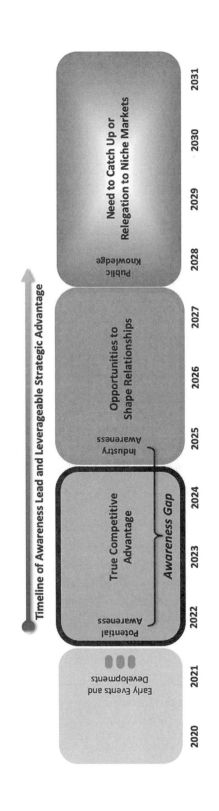

Figure 4: Bridging the Awareness Gap

the game. You will have the chance to plot your pathway to the future, perhaps even the luxury to shape the future. Later you will still have an awareness lead over most of the marketplace. But you're not ahead of your industry anymore. Then again, you're still not behind. You can still find a preferable position for your company within the changing future; plotting your pathway might be difficult now, although at least you didn't miss the boat. Without scanning, you will wake up one morning and realize that the world has moved on. Without scanning, you might end up stranded in a corner of the market that nobody else wants. Now you're in trouble.

The awareness gap is the period between the moment you could know about change and the moment everybody knows about it. Manage this gap judiciously; it is your ticket to the future. Time is of the essence; any advantage you might have dwindles over time the way sand runs through an hourglass.

Manage the Gap's Competitive Advantage

We looked at the impact music-file distribution had on the industry—and well beyond. Now we understand how scanning finds the strands from the future that alert you to new development. Another example illustrates the impact such understanding provides—now that we know of the competitive advantage that comes with early awareness. Let's look at another type of file sharing—this time in visual media.

Predating a time when smartphones were ubiquitous, February 2004's *Conversing with Images* looked at the changes cell phones with cameras could trigger in communications. The marriage of cameras and cell-phone connectivity should prove powerful. "Camera phones are potentially good news not only for handset makers but also for cellcos (mobile network operators) because of the potential revenue that consumers' picture messaging could generate," the report stated. Correct. But consumers still had to become familiar with their newfound abilities; again, the future needs time to build up its potential. "Unfamiliarity with interacting visually is evident in the messages that cell-phone subscribers send today—mostly pictures of people's faces, not of other subjects that convey a particular message. And, unfortunately, few subscribers are even sending pictures of their faces." Worthwhile mentioning: "pictures of their faces" nowadays are "selfies"; the term became commonplace only after the report's release.

From today's point of view, the author's discussion almost seems prophetic. "Subscribers (initially young or creative users) learn that visual messages constitute a communication medium that is superior to voice or text in many applications, and their use of picture messaging gradually diffuses through society."[15] Then, in January 2008, I stated in *Leveraging Photography on the Web*, "A vast repository of photos on the Web has become an enabling technology for any of a variety of purposes, including Web 2.0 applications."[16] In fact, images on the internet created an entire industry.

In October 2010, Instagram introduced its photo-sharing and posting service. Two years later, Facebook—itself a provider of a social network that heavily relies on visual media—acquired the startup. Snap's Snapchat image-sharing service went online in September 2011. Not even a year later, Vine (now Twitter) introduced a similar service that focused on six-second-long movies to share amusing or thought-provoking visual anecdotes. Communication changed (Timeline 3). The way people expressed themselves on the internet changed too. In March 2010, Pinterest launched its service as a beta version. The company allows sharing of images to create image-based portfolios of interests and hobbies—"a vast repository of photos"—that users can share online with a like-minded community.

Again, scanning offered very early awareness of how consumers will change their communication behavior—only businesses had to grab the new capabilities and package them into products and services. The awareness gap translates into competitive advantages—an early understanding of changes more than half a decade ahead of transformation of an industry . . . society, really.

Related startups became money machines: Pinterest and Snap are now (at the end of 2020) worth about $40 billion and $80 billion, respectively. And the latest incarnation of social media focused on visual media is ByteDance's by-now-notorious TikTok. Video-sharing service TikTok launched in 2017 outside its initial market of China. No doubt, the concept of visuals-based sharing and communication is alive and evolving.

But the impact goes far beyond such direct service-provision considerations. The industry features new dynamics with new players emerging and existing players snapping up startups. Marketing efforts changed dramatically with influencers and so-called Instagram traps changing brand communication, event

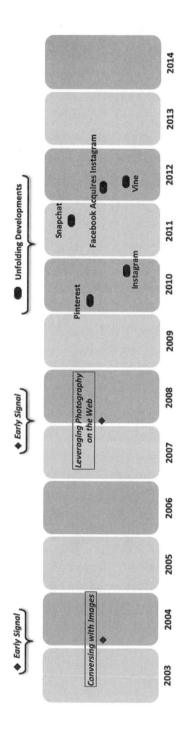

Timeline 3: Photo Sharing and Communication

design, and museum displays. And device manufacturers fed the communication frenzy with smartphone features.

Scanning's FIPI can make you live in the future. *Filter* and *identify* make you aware of changes. *Prioritize* and *initiate* prepare you to leverage your competitive advantage. Scanning works.

Setting Expectations in a Nutshell

The future is indeed uncertain. In fact, uncertainty is the only certain feature of the future. Uncertainty is a reality that requires explicit acknowledgment and consideration.

In complicated situations, advanced mathematical concepts provide answers. In complex environments, uncertainty is the defining consideration; only awareness provides guidance.

Scanning addresses developments outside one's industry, regions, and markets—areas that decision-makers are least familiar with.

Pretrend topics and *friction areas* are the ingredients of future's uncertainty. Don't push them aside; embrace them. *Pretrend topics* and *friction areas* point to changes in the marketplace.

Establish awareness early. Leverage awareness gaps to the fullest. Waiting reduces your competitive advantage. Lack of awareness might put you out of business.

Scanning's FIPI gives you a jump on tomorrow today: filtering and identifying issues create awareness; prioritizing and initiating follow-up steps enable preparedness.

CHAPTER FOUR

Filter Information

SEPARATING INFORMATION FROM DATA

---■---

The value of experience is not in seeing much, but in seeing wisely.

—WILLIAM OSLER

You hear the word *disruption* a lot these days, but what does it mean exactly? Most businesspeople think of it as transformations and discontinuities that affect businesses. Some dismiss it as the latest cliché.

Whatever your definition of the phenomenon, Netflix disrupted an entire industry with a service that its rivals did not at first take too seriously. The company went on to become a major provider of original content streamed to eager consumers. Light-emitting diodes (LEDs), a transformative technology, put the incandescent light bulb in the dark, although it took time (and regulatory boosts) for it to stride into the spotlight. Wikipedia forever changed the nature of accrued information, relying on crowdsourcing to generate continually updated content.

Scanning offers a way to see such changes coming by pinpointing pretrend topics and friction areas. *Pretrend topics* include all of the budding events and developments that could blossom into full-blown trends that will change the path to the future. *Friction areas* involve all of the little observable clues that reveal the ways emerging topics rub against existing or competing solutions and conventional or alternative business conduct in ways that can dramatically alter markets.

How do decision-makers spot pretrend topics and friction points in the avalanche of data that threatens to bury meaningful insights? They need a big bucket to hold many diverse developments, plus a fine strainer to sort out the crucial information. Scanning combines the right funnel with the right strainer.

You also need to know what not to include in your inventory of meaningful events. Fake news can distract you. Keep your eyes focused on three key sources that influence, reinforce, and can even conflict with one another: *science and technology, consumers and society*, and *commerce and competition*—the value-creation maelstrom of markets. Only by taking a holistic look at all three areas can you gain a strong sense of changes coming your way.

What types of developments foreshadow the future? To put it simply, they are little *oddities* that seem different, novel, and unusual. To put it more scientifically, they include *faint indications, discontinuities, disruptions, transformative events, inflection points, outliers*, and *unconventional wisdom*.

The Brushstrokes
to Paint the Future

Finding the information you are looking for or the information you need—the two might not be the same—requires one thing: an open mind. Information that caters to your preconceived notions of how the world works is one thing only: dangerous. Organizations love to reinforce already-existing beliefs. All is on track, we're doing the right thing, nobody's to blame. It shouldn't be a surprise that such a dynamic is all too common in successful companies. Startups question themselves; they often don't know what they are doing. Successful incumbents can look at their track record to prove that they are on track. They still might not know what they're doing.

Try to get a full view. Be warned: it is difficult. Aguilar distinguishes different types of external strategic information: "external strategic information that a manager receives," "external strategic information that a manager wants," and "external strategic information that a manager needs."[1] Scanning is here to give you what you need. You might not want it. Keep an open mind.

Much of the information you need will be outside of your market, outside of your industry. That's what makes it so difficult to get a hold on. And such information can seem tangentially associated with your business at best. Seriously . . . keep an open mind. The European Foundation highlights dangers of ignoring developments that do not directly relate to your current business. "Such developments may not be crucially important to one's immediate prospects. But if they are not taken into account until the problems start to be highly manifest, then it may be too late to adapt effectively, or the costs of coping with change may be higher than they would be otherwise."[2] Such developments build up over time; they fester. And we all know what happens when something has a chance to fester for too long.

If you go on and ignore such outside information, then you are "leaving half the business to fate," as Schoemaker warns. FIPI starts with filtering data points for manageability. Your bucket should be large. But if you take no care here, the avalanche of information will quash you. Start your four-step process with judicious information collection. Too much will overwhelm you; too little will kill you. But keep an open mind. Garbage in, garbage out is a real concern. The first steps are crucial to understand your company's outside world. After all, "Managers who can get better at managing this external uncertainty—learning how to protect against the downside and position for the upside—can start to harvest that half of the company's value that is otherwise left to the whim of the environment."[3]

Filter Today to Focus on Tomorrow

The obvious question is: where to start? In the past, getting information was difficult. Today, leaving out irrelevant information is the challenge. Many sources are obvious, and you likely already monitor them. Industry news and competitors' press releases are like breathing air. Other sources will seem odd in a professional context. But YouTube postings or private observations have value. Some sources have a reputation that makes them trustworthy; other sources require additional work to verify their credibility. Fake news has become a major concern. But fake news is not new. And not all is disinformation. More often misinformation is an issue. People make mistakes; journalists misunderstand. Stay alert!

Casting a wide web when fishing for information is crucial though. You want to see broadly. You want to capture what's brewing outside of your limited world. Don't be dissuaded when starting. It takes some time to look into nooks and crannies you never considered. Take Day and Schoemaker's comments as challenge and warning alike: "The periphery, by its nature, lacks clarity. It is uncertain and capricious. [And] In a highly connected world, faint stirrings can have large repercussions."[4] It's the lack of structure of information and the fuzziness of many events' meaning that makes this step—filtering of data—so daunting. But futurist Amy Webb comforts us: "We can observe probable future threads in the present, as they are being woven."[5]

Let's start with the process (Figure 5). Let's embark on our journey into the future. Many issues play a role when collecting data. Two considerations are crucial. They are crucial in any intelligence-gathering process. They are pivotal in scanning. Remember, we're looking for small data. There's not a lot of room for error; there shouldn't be patience for the wrong type of information. First, make sure you know what you are filtering for: accurate and relevant information. Second, make sure you understand where you get your data from: credible and diverse sources.

Find Narrative Power

Where to start is the question that is important. What can provide a signal of the future? What brushstrokes do you need to paint the picture? What tiles do you look for to complete the mosaic? There is a set of events and developments that are seeds for your narrative of the future. Many of them will surprise you. Putting them together can enlighten you—or shock you into action. Rarely will these events by themselves change the fabric of markets. (Although this is not to say that such events don't exist.) Put some of them in relationship, develop context, tell a story, and you will see how the future materializes in front of your eyes. Scanning's strength is to consider a multitude of such small changes to find the visceral dynamics that reinforce or fight each other, to provide a visceral feeling of what a changed world might look like.

Some types of events and developments will contribute to your story about the future (Figure 6). Definitions vary (academics can get into that sort of

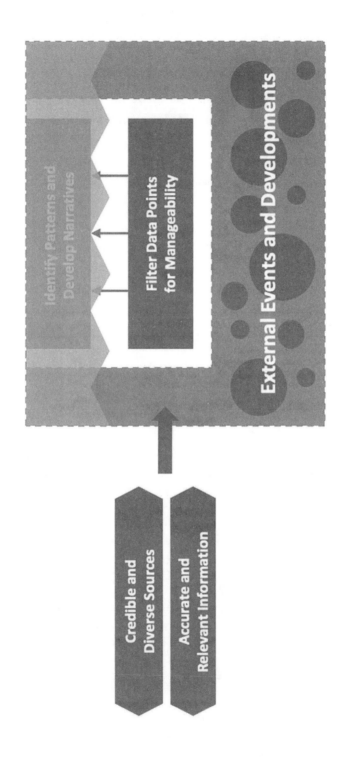

Figure 5: Step 1 Provides Relevant Input for Scanning

External/No Corporate Influence

External Events and Developments

Identify Patterns and Develop Narratives

Filter Data Points for Manageability

Credible and Diverse Sources

Accurate and Relevant Information

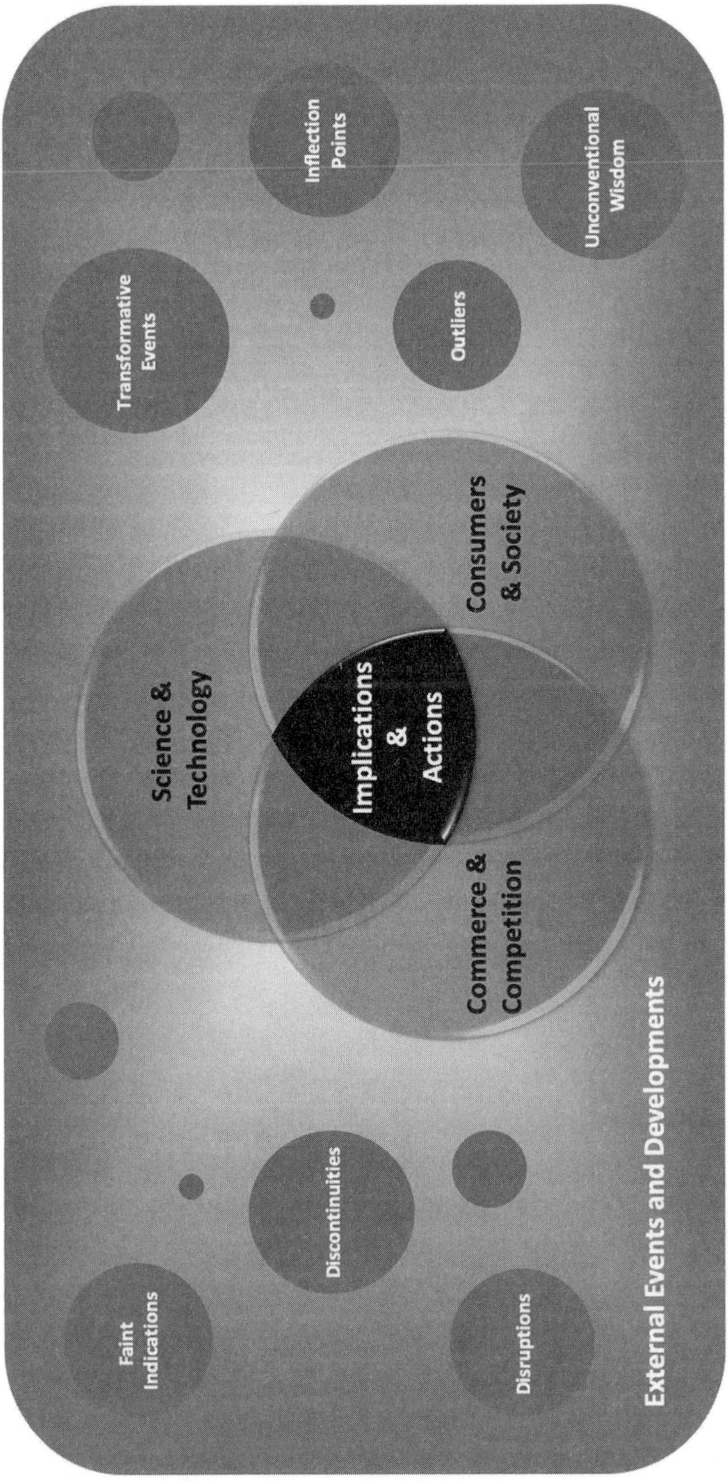

Figure 6: Developments of Interest

thing). They overlap, but that makes them rich. They are not mutually exclusive; they don't need to be. But you should know their nature and how they differ. It is what you are looking for, the filter you will apply when looking at today's world. They are the red reveal that bring the future out of a mess of data points. They are the pointers that let you see shapes and silhouettes in the blizzard of data.

So, here they are. Keep them in mind when spying into the world around you. You want to take note when you find such events and developments. Collect them. They are your brushstrokes and tiles.

Look at Early Changes

Faint indications are vague developments. They leave a lot of room for personal opinions and perceptions. They are difficult to put a finger on. But you will know them when you see them. However, be careful because they can be an excuse to include a whole bunch of meaningless events.

Timing is an important consideration in judging when considering faint indications. It is easy to be too early, running ahead of developments. Decades ago, the mobile lifestyle was such a faint indication. However, wearables and wireless networks were technologies of the future if not of science fiction novels. But Sony's Walkman gave us a glimpse into such future possibilities. The portable cassette player became available in 1979 and a runaway success in the 1980s; some 200 million units sold over the next two decades (and these players were expensive at first—really expensive). In 1984, the company introduced the Discman—same portability just for compact discs. In 1980, the indication wasn't faint anymore, but the true extent of what mobile lifestyle means took quite some time to emerge—even for advanced file formats, let alone gaming, videos, and mobile commerce. Miniaturization and increased storage were the competitive elements from that point on. Sony's own introduction of the MiniDisc in 1992 failed to excite the market. The need for a format change was palpable; only that technology choice and timing remained unclear.

In 1993, engineers at the Fraunhofer Society developed a new format for audio files: MP3. Still, the time wasn't ripe for mobile applications. In 1998, Rio, a brand of digital music players that used MP3 music files, paved the way for a lifestyle shake-up. Haven't heard of the brand? Well, timing was not on its side. Then Apple—although not the behemoth it is today—wielded its marketing heft and excited its loyal following. In 2001, the iPod line of MP3 players was born. Meanwhile SONICblue (the successor to Rio) filed for bankruptcy protection in 2003. The difference, among many things, was when iTunes and the iTunes Store entered the market in 2001 and 2003, respectively. The file management software and online file store supported iPods' usability. No, it really provided a backdrop for the mobile lifestyle. No wonder really that MP3 players aren't products anymore but mere applications of mobile devices.

Discontinuities, as the word implies, discontinue previously adopted pathways of a technology, practice, or approach. They introduce successful alternatives. They are easier to spot than faint indications, they are fairly straightforward to describe, and implications can be obvious. Because they change or eliminate current practices, they introduce friction between market players. Who will be able to leverage them therefore is unclear. But if you see them early, you have time to think about implications. Anticipating discontinuities' implications early is the success factor.

In photography, an engineer at Eastman Kodak developed a digital still camera in 1975. Only a few knew. If more had known, most wouldn't have cared. The device was too crude. A heft of 8 pounds made it more of a workout weight than a mobile device. A resolution of 10,000 pixels didn't provide beautiful renderings of the world. Oh, and it took more than 20 seconds to capture an image. No wonder then that commercial relevance was still decades away. The first commercial product was introduced in 1990. But digital photography's impact and implications emerged only in the early 2000s. Surprisingly—or rather not, given the history of companies' responses to changes in the marketplace—the company hurt most was Eastman Kodak. Wait, the Eastman Kodak that developed the technology? Indeed, and it filed for bankruptcy protection in 2012 and has since been only a shadow of its former self. The problem here is that digital photography is digital. It discontinued the need for chemicals to take pictures and movies. And Eastman Kodak was a chemical company for all intents and purposes. The story goes that Eastman Kodak didn't see the end coming. The truth is that it might have been late to think through all the implications but simply couldn't get itself to consider changing its path. (Although the company marketed digital cameras—halfheartedly.) Giving up on your core business—more like your operation's heart—is painful.

Disruptions ripple through industries and marketplaces. Disruptions question best practices and lifelong experiences. They change dynamics and necessities. But wait, isn't this entire book about disruptions? True. *Disruption* has become a synonym for drastic change. So, these changes include discontinuities and transformations, for instance. But then there's the initial meaning, which is much narrower. And it's worth keeping these definitions separately in mind— because a true disruption is quite an insidious beast. You likely don't see it coming. If you do, you discount it. But once it unfolds its power, you're toast more likely than not. True disruptions do not occur very often.

The concept of disruptions emerged from the research of late Harvard Business School scholar Clayton Christensen.[6] Christensen himself took offense at the overuse of the term, but what can you do? "Despite broad dissemination, the theory's core concepts have been widely misunderstood and its basic tenets frequently misapplied." Genuine disruptions start out in market corners that incumbents don't particularly value. They hit you as they unfold their true power when it is too late for you to respond in a meaningful way. Christensen and colleagues argue, "Many researchers, writers, and consultants use 'disruptive innovation' to describe any situation in which an industry is shaken up and previously successful incumbents stumble. But that's much too broad a usage." Here's what makes such developments so sneaky. "'Disruption' describes a process whereby a smaller company with fewer resources is able to successfully challenge established incumbent businesses. . . . Entrants that prove disruptive begin by successfully targeting . . . overlooked segments, gaining a foothold by delivering more-suitable functionality—frequently at a lower price."[7] A disruption starts with something you, the big player, don't care about.

Netflix's rise to success is a perfect example of a true disruption. Netflix, the movie rental company, initially targeted a very different segment than traditional movie-rental companies served. Blockbuster and Hollywood Video focused on consumers who wanted the newest releases and catered to impulse. How does Netflix not cater to that group? Well, when it started out in 1997, it required customers to select DVDs online and organize them in a list. The DVDs then were sent via mail—snail mail, that is. But first you had to return your previous rentals via mail before receiving new ones. The process took up to five days. From today's point of view, five days is an eternity. But even two decades ago, Netflix couldn't exactly cater to Friday night spontaneity. By the time Netflix shifted to streaming applications, the traditional players were ambushed by a player they had laughed at before; both companies went out of the rental market in 2010. Neither of those companies exists anymore. Painful, the unworthy player put a knife into the heart of decade-long success stories. And here's the scare: obvious competitors—even with vastly superior offers—do not represent Christensen's original definition of disruption. You know them; you pay attention. But the inferior players on the sideline can be the center of future change—only you don't see them or, worse, dismiss them. "The fact that disruption can take time helps to explain why incumbents frequently overlook disrupters"—you definitely want to make them part of the data you collect.[8]

Keep an Eye on Crucial Developments

Transformative events can be fairly easy to spot. Then again, the world's complexities and your distraction can hide them. In contrast to disruptions, transformative events address a need or problem with a better solution. The new approach directly competes with incumbents' products and services. Silicon Valley certainly has shown the transformative power the integrated circuit had on vacuum tubes. And yes, chips went on to change pretty much everything else. They were better than what came before; they didn't necessarily change the concept. The impact was all enveloping though. Transformative developments don't have to be earth shattering across all commercial spaces. But if your business is affected, it won't stay the same for long.

Enablers trigger transformations. Unsurprisingly, the internet is now the grand information and knowledge repository. Surprisingly, though, for the longest time the market for encyclopedias (printed or digitally) remained fairly intact nevertheless. It took until 2001 when Wikipedia emerged. The "web-based, free-content encyclopedia" transformed the business.[9] It then took until 2010 until the most prominent example, the grande dame of encyclopedias—the Encyclopædia Britannica—published its final printed volume after more than two centuries of publications. (Microsoft Encarta, a digital encyclopedia, ceased to exist the year before.) Don't think the transformation is over. Currently, most companies in virtually all industries struggle to adapt to the transformative power of digitalization. (The Internet of Things will continue the transformative power of the network.) Here's the thing about transformative technologies. Amara's law states, "We tend to overestimate the effect of a technology in the short run and underestimate the effect in the long run." (Roy Charles Amara was a research fellow at the Stanford Research Institute before it changed its name to SRI International.) The statement should come as a red flag for decision-makers. Bazerman and Watkins describe the same sentiment from a business perspective: "Research studies consistently turn up such myopic preferences, which seem to reflect an extremely high discounting of the future. That is, rather than consciously evaluating options from a long-term perspective, people tend to focus on short-term considerations."[10] And that is why you are reading this book, why you are trying to make looking in the future a habit.

Inflection points are mathematically defined as the change of a plotted curve. Here you have it. For all other purposes, an inflection point is when the number of sales or users takes off. It's a turning point, the point where a particular application finds exploding growth. Unfortunately, it's also the point where sales start to taper off. Inflection points are crucial markers in trend lines. They are very easy to spot in hindsight. They are devilishly difficult to pinpoint when they actually happen. And that is the moment when they matter because implications abound. Sales growth that starts to taper off is a warning sign for companies to lower manufacturing capabilities and production targets. Perhaps you even want to consider exiting the market? Here's the rub though. Sales growth tapers off, but number of sold products still increases during that time. They just sell at a decreasing growth rate. The inflection point can be the point where sales reach a plateau—or start to go downhill. Now you see how easy it is to miss them or misinterpret them. Missing an inflection point that suggests declining sales while adding manufacturing capabilities and enlarging distribution networks is disastrous.

It's also easy to see how major markets can be missed. Mainframe computers—large furniture-sized computers that were stored in dedicated rooms and required dedicated personnel—entered the business world in the late 1950s. Besides a limited number of such rooms, the world remained analog. It perhaps should be no surprise that the impact of this new technology was easy to underestimate at an early stage. Famously, Thomas Watson, the president of IBM—a manufacturer of business equipment that would go on to become a major player in computing—stated in 1943, "I think there is a world market for maybe five computers." Desktop and personal computers changed the technology's visibility starting in the 1970s. Apple's Apple II and Commodore's PET, both introduced in 1977, made computing available to individuals, and 1982's Commodore 64—finally brought computing into the developed world's living rooms (or kids' rooms, that is, in many cases). Life changed as computing power democratized—and made an entire generation of youngsters a future work pool of software developers. Tellingly, in the very same 1977, Ken Olsen, founder of Digital Equipment Corporation, proclaimed, "There is no reason anyone would want a computer in their home." Standing at the verge of an inflection point does not have to be obvious.

Identify Unusual Events

Outliers also stem from a statistical observation but find use in scanning in a much broader context. For our purposes, outliers in business models, consumer behavior, or entrepreneurial conduct are outside of common and understood practices. Perhaps a business owner is inexperienced and makes mistakes. Or a small group of consumers is truly idiosyncratic and their behavior does not apply to the rest of the market—the oddballs at the fringe. Perhaps. Or it's the beginning of a new chapter in economic activities. Outliers are very different from common conduct and historical experience. They can be the result of, for instance, an entrepreneur's incompetence and lack of knowledge—and therefore will result in the business's failure. Or an entrepreneur finds a new way to look at the market—and changes the understanding of business practices in the process. Outliers can be as much distracting red herrings (don't worry; the scanning process will take care of it) as they can be valuable early signs of change. Dismissing them comes at a peril though.

Anecdotal evidence illustrates the issue. Fred Smith, a student at Yale University, developed an idea for a reliable overnight-delivery system for urgent shipments. According to lore, he submitted a paper in class in the 1960s that outlined a hub-and-spoke system of air transportation and truck fleets that would enable such a system. His professor didn't see the merit. In the 1970s, Smith went on to found FedEx with overwhelming success. (Smith indeed submitted the paper but couldn't recall the actual grade, instead mentioning to a reporter, "I guess I got my usual gentlemanly C," giving rise to the anecdote.[11]) Whatever the real story, the anecdote stuck and certainly serves the purpose of being a reminder that truly new concepts and approaches—outliers—can change the business environment. It also shows that even trained observers can have a hard time seeing their value.

Unconventional wisdom represents approaches and conduct that do not conform to accepted standards and developed practices. Unconventional wisdom shares similarities with outliers. Unconventional wisdom embraces working with untested approaches and novel strategies to address market needs from different directions. Such revolutionary thinking can open very new commercial pathways. Or it can become a flash in the pan.

The dot-com period from 1997 to 2000 is a high-profile example illustrating how unconventional wisdom can first change the approach of an entire industry. Then it turned into an example of how an industry can outrun its headlights. Internet-based companies experimented with a range of novel approaches and theories during that time. Then an increasing number of entrepreneurs jumped on the bandwagon. The belief in the power of a "New Economy" attracted investors like a flame is irresistible for moths. The new economy created its own vocabulary: considering network effects, attracting "eyeballs," and developing mindshare led to calls for "growth over profits" and "get large or get lost"—the idea that very rapid growth in audience and market attention was more valuable than revenue generation and profits. In hindsight such thinking clearly seems to have obvious limitations; at the time, excitement trumped rationales. Most strikingly looking back was the attention to a new metric: the burn rate, a measure for the speed at which startups were using up investors' capital. Counterintuitively from today's—or any other time's—point of view, many observers perceived a high burn rate as a proxy for rapid business development and promising prospects—undeniably a form of unconventional wisdom. Exuberant spending on launch parties to create awareness became a frequent occurrence. "Prefix investing" described the tongue-in-cheek practice of gobbling up shares of companies that used an e-prefix for *electronic* with the hope that internet-based approaches would automatically yield high return rates.

Beginning in 2000, the infatuation with this type of unconventional wisdom came to a sudden stop, and the NASDAQ Composite index (a stock index that skews toward information-technology companies) peaked at more than 5100 points in March 2000 and then rapidly declined to a final low of just barely above 1100 points in October 2002 (it took more than twelve years then to move beyond 5000 points again). This episode shows how ambiguous unconventional wisdom can be. Whereas pet-supply retailer Pets.com and online grocer Webvan have become cautionary tales, Amazon and Google have moved on to become commercial juggernauts. The unconventional wisdom represented a bundle of short-term fads and genuine commercial shifts. Unconventional wisdom can be exciting. Caution is advised; conventional wisdom often exists for a reason.

The impact of this time's exuberance and failures is still palpable. Many concepts proved prophetic. Some unconventional wisdom won over critics and doubters. Network effects matter. In fact, many analysts now argue that the hurdles for entry these effects unleash resemble monopolistic power; the call for regulations is getting louder. The concept of the *long tail* changed retail. In physical stores, the focus is on selected few titles that sell in large volumes. Web-based shop fronts with back-end warehouses can also efficiently serve markets with many titles that sell only in small numbers. The concept gained business popularity after author Chris Anderson highlighted its relevance for ecommerce in 2004. Even very rarely ordered products can be kept at a central warehouse; there is no need to provide each branch with a unit. Ecommerce was able to overcome the "tyranny of lowest-common-denominator fare."[12]

The different types of event categories deserve distinction. Understanding what makes them unique will help your narrative. Many discussions are awash in hyperbole. Such discussions lose not only credibility but also narrative power. If every event out of the ordinary readily is considered a sea change, a game changer, or a paradigmatic shift, the sea in reality has no shape, the games no rules, and the meaning of paradigms ceases to exist. Words describe meaning; hyperbole blurs relevance. Descriptions and categories convey understanding; words matter.

In the end what you're looking for are the pieces of information that surprise. The brushstrokes you don't expect, the tiles that don't fit in your assumptions. That's because they will paint the picture, create the mosaic that tells you something new about the future. These are the pieces that make you aware of emerging worlds.

The Stones to Turn

Now the question is: Where should you look for information? The easy answer is: wherever you find it. Casting a wide net to access information from as wide a set of sources as possible will provide you with a reliable stream of information. The challenge is a formidable one. You want to find information, but you don't want to be overwhelmed. You want to make sense of the future, but the bits and pieces of information don't tell a story until you put them together. Aguilar underscores the difficulties associated with "the sheer weight and volume of relevant data, including the disorganized, fragmentary and unverifiable state in which the data are ordinarily found."[13]

Let's start with where to find such information before we try to organize it. There are some limitations—not surprisingly. Where to filter strategically relevant information from always has played a role to ensure quality standards. With the advent of so-called fake news—the deliberate creation and dissemination of misinformation and false assertions—arguably where to filter from, the source of the information, has become a consideration as important as what to filter for, the content of the information.

Factual information has come under such attack that Oxford Dictionaries declared *post-truth*—"relating to or denoting circumstances in which objective facts are less influential in shaping public opinion than appeals to emotion and personal belief"—the word of the year in 2016.[14] Post-truth and fake news make a deliberate selection of sources that find use in scanning a vital quality requirement.

The Usual Suspects

Primary sources. Leveraging existing and ongoing research and operations should come as a natural starting point. Similarly, ongoing operational activities should provide information that can find use. Activate your team members. Internationally, employees and business partners tend to have a sense of the respective countries' conditions and economic developments. On-the-ground observers can play an invaluable role in sensing developments early. There are obvious challenges to keeping up—let alone establishing or activating—such a network. Look at your partners. Customers and suppliers should even be considered as participants in your scanning efforts.

Institutions and organizations. Besides commercial players, a wide range of organizations exist to do research, advocate causes, represent groups within societies and economies, or drive public opinion and strive to progress societal considerations. Universities, R&D institutes, governmental departments and institutions, nongovernmental organizations, think tanks, and consulting groups all can play a role. Academic journals should be a constant source of information. Many technologies and new understanding emerge in academia. Use such sources; don't get lost in details. And many countries have consumer-advocacy organizations that provide information and ratings on markets, products, and services.

Patents. Many companies investigate patent filings and grants in detail to understand future technological environments. A research community out there is focused on this type of information. A selected few patent filings can be of utmost importance and deserve scrutiny, but instead most join a sea of patents with limited commercial significance. They can distract.

Traditional media. Cast a wide net; avoid dubious sources. Broadcast media, television and radio, can very conveniently provide valuable information. The general nature of the media also offers a framework to gauge public relevance, directions policy makers want to pursue, and high-level commercial activities. Print media offer an even wider range of sources to consider. Online publications of magazines and newspapers have brought the discovery of crucial changes in the commercial and societal world within a click's reach—literally. Business media, newspaper, and magazines, can provide obvious advantages for looking at commercial developments and considerations. These media outlets look at technologies, society, and market dynamics through a commercial lens. International media are crucial to add regional breadth and diversity to the mixture. Consumer-focused media can offer the customers' perspective. More specifically addressing consumer interests in emerging technology areas, hobby and DIY publications offer insights into the way consumers adjust their environment to fulfill their unmet needs.

Some articles can point to emerging developments convincingly. We previously discussed Chris Anderson's 2004 article "The Long Tail."[15] His commentary in "The End of Theory" in 2008 spelled out some aspects of big data applications that since arguably have seen wide discussion and interpretation.[16] Both articles, and indeed their titles, have become shortcuts to reference related concepts and possibilities. Similarly, in 2008 Alex "Sandy" Pentland, professor at the Massachusetts Institute of Technology, presented the concept of "reality mining." His academic work allowed a much broader understanding of how future data applications could extract valuable commercial data. *Reality mining* "is all about paying attention to patterns in life and using that information to help [with] things like setting privacy patterns, sharing things with people, notifying people—basically, to help you live your life."[17] It is not difficult to see that Pentland early on put concepts such as self-tracking, quantified self, big data, privacy management, and digital assistants into a larger framework.

Books on current topics can set agendas and drive commercial activities. Books, despite their longer publishing cycle, can shine a light on emerging issues. On October 27, 2015, Ted Koppel's book *Lights Out: A Cyberattack, A Nation Unprepared, Surviving the Aftermath* was published.[18] Koppel looked at the US energy infrastructure and found it utterly inadequately prepared to defend serious hacking attacks. The call came in time—without heeding though. Two months later, on December 23, 2015, Russia launched a cyberattack on Ukraine's energy network; this example is considered the first known attack of its type. What about the US then? Well, in October 2017, the US Federal Bureau of Investigation and Department of Homeland Security issued joint Technical Alert TA17-293A—a warning to energy organizations that describes an advanced and persistent threat to the energy sector. "Since at least May 2017, threat actors have targeted government entities and the energy, water, aviation, nuclear, and critical manufacturing sectors, and, in some cases, have leveraged their capabilities to compromise victims' networks."[19] Technical Alert TA18-074A, from March 15, 2018, "provides information on Russian government actions targeting U.S. Government entities as well as organizations in the energy, nuclear . . . sectors."[20] Although books can't be as current as articles, they can provide you with comprehensive glimpses into the future.

New media. Another source of information comes from new media—a term that seems to have only historical relevance now—media that is internet-based and available on demand. Most newspapers, magazines, and academic publications, but not all, have an online outlet. Use social media to get a feeling for the landscape of consumers and professionals. Social media provides a wealth of information, but the majority of the information is disorganized trivia of questionable authenticity and substance. Social media is also prone to false evidence and exaggeration. Be careful! Recent events highlight the ability of their use as propaganda.

Conferences, trade shows, and scientific gatherings. Events and meetings provide another crucial input source for scanning efforts. They can offer an overview of new developments. After all, these events try to show off new research, concepts, products, and services. Attendees here very easily develop networks of like-minded people. Build such networks; leverage their knowledge.

Technical demonstrations tend to be marketing affairs. But sometimes they foretell the future in almost eerie ways. In perhaps the most famous example of such demos, on December 9, 1968, Douglas Engelbart—researching information technology at Stanford University's Stanford Research Institute (now SRI International)—gave a live demonstration of new applications his team developed. The demonstration, which later became known as the "Mother of All Demos," introduced collaborative-work protocols, the computer mouse, hypertext, video conferencing, and window applications, among other features. The demonstration preceded (and highly influenced) work at Xerox PARC, which itself provided the crucial impetus to Apple's and Microsoft's focus on graphical user interfaces. Other aspects of the demonstration had a very direct effect on the development and usability of the World Wide Web. In other words, a presentation that lasted merely ninety minutes at the end of the 1960s provided a glimpse into the computing world of the 1980s and 1990s and beyond. (The demo is available online; watching it with today's knowledge of what the world looks like in 2020 makes the demo not a memo but a manuscript from the future—worth revisiting.)

The Unusual Suspects

Individuals of public interest. Some business personalities can become pundits in their own right. The opinions of business leaders with widely accepted knowledge in a field and a proven track record in their industry have to find consideration; for one, many people will heed their advice and take their comments at face value and thereby change the fabric of markets by their own actions. Another aspect is worth taking into account: Warren Buffet, the chairman and CEO of Berkshire Hathaway, and Tim Cook, Steve Jobs's successor at Apple, have very substantial amounts of money at their disposal. Mere curiosity about a market opportunity can shift hundreds of millions of dollars, whereas true interest in related opportunities can mobilize billions of dollars.

Other individual businesspeople have become celebrities. They themselves are the events you're trying to capture. Sir Richard Branson has attracted quite a lot of attention since he started his run through industries with his Virgin Group and subsidiaries. Elon Musk has shown a similar ability to move seemingly seamlessly across industries with his various business ventures. Arguably, he single-handedly (well, with help of collaborators) changed the online payment world, the automotive industry, and even space's commercial prospects. Similarly, Jeff Bezos and his innovation vehicle Amazon have shown the ability to repeatedly shake up the market landscape of online and recently brick-and-mortar retail, with consumer electronics and even entire business-to-business services. His company's R&D subsidiaries, A9.com and Lab126, investigate innovative technologies and novel consumer electronics.

Also, *opinions* matter. They do pose some challenges though; strictly speaking, they are not factual data points and clearly feature bias and political leanings. But opinions matter. "Musk argues. . ." or "Bezos believes. . ." captures potential changes: the messenger can become the message. Opinions matter because they can set agendas and outline expectations in an industry. Moore's law was initially an observation, an opinion, presented by Gordon Moore of Intel; successively, it became a guardrail for an industry roadmap. Opinions matter because experts in a field provide a point of view and a perspective of value. When Apple introduced its music-streaming service in June 2015 to compete with successful existing services, musician Taylor Swift used her Tumblr social-media account to publish an open letter to Apple—which has become known as Ms. Swift's love letter because of the letter's title "To Apple, Love Taylor"—explaining that the company's royalty policy leads her to withhold her newest album—*1989* at the time—from the service. "We don't ask you for free iPhones. Please don't ask us to provide you with our music for no compensation."[21] In reaction Apple changed its policy, and the financial conflicts between artists, media companies, and consumers became a more prominent discussion point. Ms. Taylor's opinion mattered, triggering commercial change.

Finally, one's own *personal observations* can round off information. Discussions and communication within professional networks offer an opportunity to pick up developments earlier than any media source can. Similarly, personal observation in day-to-day life, even in an individual's own home environment, can offer an opportunity to identify developments earlier than many organizations and research institutions can; your children can tell you a lot about societal changes. Your own changes in the way you use technology can indicate more general consumer changes.

Similarly, observations of your daily environments should be fair game as an information source. For example, look at the way tattoos and some types of body modification have spread from niche behavior to mainstream behavior. Niche behavior now includes a very small group of people that self-embed RFID chips or sensors as biometric security devices or for connectivity and quantified-self applications. Users of *wetware*—permanently embedded technology-based interfaces—represent a very defined niche market. However, this small market could offer an early understanding of consumers' willingness to adopt human-augmentation technologies in the future.

Personal observations, per definition, are subjective and live on anecdotal samples; they therefore require particularly careful curation as data sources. But personal observations also can offer a very unadulterated look at emerging developments. Use such insights . . . carefully, but use them.

When I was traveling to Venezuela in 2008 to provide scanning services to clients, the state of the economy became very obvious very fast. The international media had some concerns, but personal experiences made the degree of the problems glaringly obvious. Many international companies' representatives very bluntly stated they did not see a future in the country; in fact, one businessman pointed to a suitcase in his office he had already packed to leave the country the very next day. Perhaps even more striking was the workshop with Petróleos de Venezuela, S.A., or PDVSA. The oil company found itself at the center of political changes because it was—still is—the country's revenue machine. The workshop was to be kept in line with the political agenda set by the government. Before the workshop I was advised that some terms—such as *profit*—would not be welcome. And some workshop participants were sent to the meeting to ensure adherence to the guidelines. Clearly, the economic state of the country was in trouble. In addition, casual dinners with business contacts and conversations with hotel and restaurant employees revealed growing concerns of economic developments. The lesson is obvious; not even a week of personal experience in a country can turn spotty media coverage into a visceral understanding of a market.

Keep in mind that it is about the narratives that you will develop. Author Martin Lindstrom explains, "In-person observation, and a preoccupation with small data is what sets apart what I do in a world preoccupied by big data." He also cautions, "A lone piece of small data is almost never meaningful enough to build a case or create a hypothesis, but blended with other insights and observations gathered from around the world, the data eventually comes together to create a solution."[22]

Finally, *anecdotes* can count. Even though Amazon is very focused on data analysis of its millions of daily online and real-world transactions to serve customers best and provide the most relevant product recommendations, its CEO Jeff Bezos is very aware of the power of small data. Speaking at the Forum on Leadership at the George W. Bush Presidential Center in April 2018, he revealed, "We have so many metrics. . . . The thing I have noticed is when the anecdotes and the data disagree, the anecdotes are usually right. There's something wrong with the way you're measuring it."[23]

Bezos is not alone. When you are dealing with a future created by new factors and dynamics, past data is an ill adviser. Andrew S. Grove, former chairman and CEO of Intel, puts a finger on the issue: "The point is, when dealing with emerging trends, you may very well have to go against rational extrapolation of data and rely instead on anecdotal observations and your instincts."[24]

As a matter of caution, don't expect the future to serve itself on a silver platter. Information can be hidden. Think in tangentials and implications. Try to look at unusual sources. Li Keqiang—Premier of the State Council of People's Republic of China—once told a US diplomat that he himself has only limited faith in his country's own GDP index and reporting. He instead trusts secondary information such as the development of railway freight, the growth of bank loans, and, in particular, electricity production to infer the overall strength of the economy. With scanning, always infer; always make connections. You will see: it can become second nature.

Filtering Information in a Nutshell

Filtering data points for manageability is a crucial enabler for developing workable patterns and narratives. Filtering allows you to focus on a manageable set of events and developments.

. . . .

External information represents value that many companies do neither judiciously capture nor appropriately process. Such information is your ticket to competitive advantages. It's your policy against surprises.

. . . .

Cast a wide net to capture the information you need; missing crucial pieces here restricts your vision. Look across geographic regions to see what changes take place in *science and technology, consumers and society,* and *commerce and competition.*

. . . .

Faint indications, discontinuities, disruptions, transformative events, inflection points, outliers, and *unconventional wisdom* are what you are looking for. Understand differences, but use their narrative powers.

. . . .

Information can be wrong, misleading, or distracting. Some sources require more careful consideration and evaluation than others. Look at the widest range of sources possible, but be aware of the provenance of your information.

. . . .

Some information sources are less objective than others. Don't discard such sources offhandedly. Anecdotal experiences, opinions, and personal observations can provide texture and insights you might not gain otherwise.

. . . .

CHAPTER FIVE

Detect Patterns

MAKING SENSE OF TODAY'S WORLD

■

*The greatest value of a picture is when it forces us to
notice what we never expected to see.*

—JOHN TUKEY

Set your time machine for 2006. In that year Muvee releases software that allows automatic editing of personal video clips. Thomson Financial starts using software to generate financial reports. And the head of the Computational Bioinformatics Laboratory at Great Britain's Imperial College proposes a future in which robotic scientists will replace some human lab assistants. In November of that year, my team put these events in context and concluded, "As automation climbs the employment value chain, traditionally secure middle-class jobs in a variety of fields are at risk."[1]

Now whisk ahead to 2011, when MIT researchers Erik Brynjolfsson and Andrew McAfee release the book *Race Against the Machine*, in which they argue, "The AI revolution is doing to white collar jobs what robotics did to blue collar jobs."[2] Two years later, McKinsey Global Institute listed automation of knowledge work as a disruptive technology. Then, in January 2018, Joseph Pistrui, professor at IE Business School in Madrid, proclaimed, "Technology is going to replace jobs, or, more precisely, the *people* holding those jobs. . . . Knowledge workers will not escape."[3]

Scanning helps sketch the robust patterns that put an abundance of data in a useful context. It provides a virtual time machine you can use to transport yourself from today's data patterns to an often-surprising future. Of course, decision-makers who propel their imaginations from today to tomorrow can begin seizing opportunities their competitors cannot even begin to imagine.

The Danger of Shiny Objects

Now that you have your data points, how do you get information from them? Many companies look at scanning as finding golden nuggets. In other words, the work is more or less done by filtering interesting points from the external world. And no doubt, some of the small data you collected might find immediate use. A new application for an existing technology, a new algorithm that can open a new market segment or complement your portfolio—they will be welcome additions to your existing strategy.

But here is the problem with this approach: it simply falls short. It ends too early. You still have not seen the future. Why stop when you just got going?

Single data points are not reports from the future—perhaps short memos distributed in the here and now. A research result can be preliminary and might require confirmation. A new prototype might be a researcher's pet project, not a market's call for a new solution. And a new startup's market approach might be the result of an entrepreneur's inexperience rather than a unique perspective on market needs.

Corporations that focus on the identification of single data points rely on finding small "nuggets of wisdom." Nuggets can be misleading. Worse, nuggets of wisdom more often than not represent wishful thinking: you found what you wanted to find.

Nuggets are important ingredients to develop a strong comprehensive narrative. But they do lack the robustness and narrative power to justify a strategy. Singular developments likely will lead you astray. A number of similar developments or approaches meanwhile indicate that there's a method to the madness. Various market participants do something similar. A pattern just formed.

Of Brushstrokes and Tiles

Patterns provide robustness; you overcome the reliance on a single data point. Patterns are your first indication that change might be occurring. They are your weak signals of change. They should wake you up. Patterns become elements of stories—narratives that allow you to immerse yourself in the future and to explore your position. Brushstrokes become pictures; tiles start to form mosaics.

Although many futurists see single data points as weak signals of change, I perceive weak signals already as a result of some preliminary work, pattern recognition. Practically, it is desirable that some thought and initial sensemaking go into the creation of signals, even if they are considered to be weak. Weak signals then are patterns, sets of relevant information pieces that feature relationships or point to novel dynamics. A true signal of change, in my understanding, is then a narrative that went through a process of vetting and additional research.

Before you can identify meaning in future worlds, you will have to go through pattern recognition and narrative development. The exercise will enable you to make sense of today's world and then extract knowledge about tomorrow's world. You just entered the second step of FIPI (Figure 7).

You have your data points—your little memos of the future. You have a process in place to monitor events and developments in *science and technology*, *consumers and society*, and *commerce and competition*. You extracted *faint indications, discontinuities, disruptions, transformative events, inflection points, outliers,* and *unconventional wisdom*. You have a small set of data points in front of you. Now you will start making sense of it.

To really understand the world of tomorrow, you will need a broad range of experts with experience and expertise across disciplines and regions. Bring them together in a workshop to make sense of it all. As they continue to work with the process, they will become comfortable in voicing unconventional concepts and out-of-the-box ideas—even the occasional outrageous thought. All good, all important. After all, "If people did not do silly things, nothing intelligent would ever get done," as philosopher Ludwig Wittgenstein found.

At this stage, you are not looking at analysis of today. You are trying to spin creative ideas about tomorrow. Let analysis follow; don't let it limit your perspective here. If workshop participants don't feel comfortable mentioning hunches,

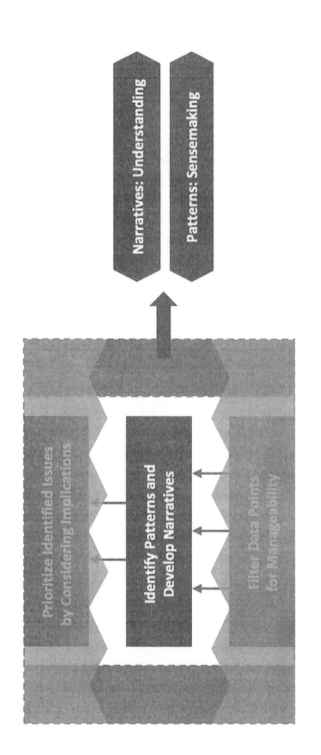

Figure 7: Step 2 Makes Information Meaningful

intuitions, and half-baked ideas, truly forward-looking perspectives will not emerge. An analytical phase afterwards can clean up arguments, tie together loose connections, and research details.

Establishing a nonthreatening atmosphere in such workshops is therefore a crucial prerequisite to gain genuine and novel insights. Problematically, such openness and freedom to explore creativity can feel very unusual for many participants. Creating openness and freedom within otherwise strictly hierarchical organizations can be a formidable challenge. Although it is tremendously beneficial to have not only a representation of different disciplines in the workshop but also participants from the organization's various hierarchical layers—from intern to strategic decision-makers—in some corporate and regional cultures the inclusion of direct supervisors can be problematic, stifling.

The payoffs of atmospheres of mutual respect and the freedom to explore new ideas can be tremendous though. Participants will willingly share their knowledge and background, and the setting can embolden them to move beyond their own comfort zone and speculate on potential developments. Idea-inviting atmospheres will encourage individuals to take a completely different look at their current knowledge and discipline. Tapping into open mindsets can provide access to issues shared within the organization but never voiced.

Tapping into these opinions also opens a door to the knowledge you already have in your organization—a door you never opened. McGrath points to such untapped insights: "I often find in firms that are surprised by an inflection that the knowledge about what was going on was actually plentiful in the organization—somewhere—but no one had a complete enough picture to make sense of it."[4] Use such meetings to get participants to share their concerns and insights to complete pictures.

Participating in such a workshop can have long-lasting, organization-wide effects. The experience can empower employees to pursue their day-to-day work with newfound meaning to their own contributions; keeping the future in mind assigns relevance to individuals' contributions.

During such meetings, patterns of today's events and developments will emerge. Patterns that directly point to future dynamics. Patterns that are weak signals of change.

Webb highlights how patterns can provide guidance and help practitioners understand plausible developments: "The future doesn't simply arrive fully formed overnight, but emerges step by step. It first appears at seemingly random points around the fringe of society, never in the mainstream. Without context, those points can appear disparate, unrelated, and hard to connect meaningfully. But over time they fit into patterns and come into focus as a full-blown trend: a convergence of multiple points that reveal a direction or tendency, a force that combines some human need and new enabling technology that will shape the future."[5]

But be careful with trends. "Trends essentially represent a noticeable shift," according to business strategist Steve Tighe.[6] They are already noticeable, nothing to anticipate here. They are the outcome of what we are looking for. Once we know there is a trend, we're in a flock with other sheep, trying to make sense of it. We're looking for pretrend topics to be ahead of the flock.

The Power of Patterns— Anticipate Tomorrow Early

Let's take a closer look at the example of automation from the beginning of this chapter. Automation is lauded as an enabler of commercial progress, and it is blamed for eliminating jobs across a wide range of industries. In the past, automation was seen as a concern for mainly blue-collar jobs. Not anymore. Increasingly, policy makers realize that white-collar jobs are now affected as well. Robotics, artificial intelligence, and software-driven applications will have wide-ranging impact on the employment landscape. However, what kind of impact is still under debate.

In 2011, Massachusetts Institute of Technology researchers Erik Brynjolfsson and Andrew McAfee released their book *Race Against the Machine*.[7] Media entrepreneur Tim O'Reilly wrote in a review at the time, "The AI revolution is doing to white collar jobs what robotics did to blue collar jobs."[8] This topic and these implications have found widespread recognition. The researchers discuss how advancing technologies are competing with an increasingly wide range of human skills and jobs. The book became a marker of a new employment threat for many industry observers. But half a decade earlier, pattern recognition found such a development. Let's take another look at the developments mentioned in this chapter's introduction to sort them into a timeline.

In 2006, indications accumulated that a new stage of automation had occurred. In discussions at that time, participants of scanning meetings pointed to a very different work environment for workers and a different division of labor inside organizations. At the time the head of the Computational Bioinformatics Laboratory at Imperial College foresaw a future in which robotic scientists would replace some human assistants in laboratories. Meanwhile, Muvee offered software that allowed automatic editing of personal video clips. And Thomson Financial, a financial information service, started using a software solution that generated financial reports within fractions of seconds after receiving data.

The year 2006 already had tidings of things to come. Yesterday did in fact let us know about today, if not tomorrow. In November 2006, in *Automation Climbing the Value Chain*, Andrew Broderick cautioned, "As automation climbs the employment value chain, traditionally secure middle-class jobs in a variety of fields are at risk."[9] Compare this statement to O'Reilly's review half a decade later. Not verbatim, but pretty close. Half a decade had to pass for the sentiment to hit the spotlight. Half a decade that you, for example, could have used to prepare yourself for a changing employment landscape. Again, ultimate implications are still debatable; the general development is not.

In the years after *Race Against the Machine*, researchers still tried to catch up with the development—a development that pattern recognition indicated in 2006. In May 2013, McKinsey Global Institute listed automation of knowledge work as a disruptive technology and ranked the development second among twelve technologies with the highest "estimated potential economic impact of technologies across . . . applications in 2025."[10] And in January 2018, Joseph Pistrui, professor of Entrepreneurial Management at IE Business School in Madrid, argued, "Technology is going to replace jobs, or, more precisely, the *people* holding those jobs,"[11] emphasizing that "knowledge workers will not escape." He tackles the question of what qualities will help human workers continue to be relevant and believes "imagination, creativity, and strategy" are skills that remain relevant in an increasingly automated business environment.

Timeline 4 illustrates how far ahead of the curve you could have been. We'll come back to this development one more time to see how we continue to pick up strands from the future—strands that tell us that this development is far from over.

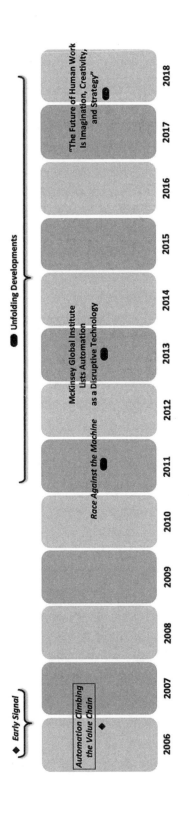

Timeline 4: Automation's Growing Capabilities

◆ Early Signal
● Unfolding Developments

Automation Climbing the Value Chain

Race Against the Machine

McKinsey Global Institute Lists Automation as a Disruptive Technology

"The Future of Human Work Is Imagination, Creativity, and Strategy"

2006 2007 2008 2009 2010 2011 2012 2013 2014 2015 2016 2017 2018

The moment when most observers became aware that automation had extended its reach beyond blue-collar work was the publication of 2011's *Race Against the Machine* and media resonance to the book that enforced the message. Brynjolfsson and McAfee warned, "We're entering unknown territory in the quest to reduce labor costs. The AI [artificial intelligence] revolution is doing to white-collar jobs what robotics did to blue-collar jobs. . . . No one else is doing serious thinking about a force that will lead to a restructuring of the economy that is more profound and far-reaching than the transition from the agricultural to the industrial age."[12] Compare the core premise to 2006's *Automation Climbing the Value Chain*: "Positions that people once considered to be relatively secure because of the higher skill levels that they require may not in fact be immune from the threat of automation. As automation climbs the employment value chain, traditionally secure middle-class jobs in sectors as diverse as scientific research, medicine, multimedia production, financial reporting, and financial services are at risk."[13] Brynjolfsson and McAfee certainly provided more detail and analysis, but the claim to providing an early alert belongs to the scanning process.

Since then, two developments have shaped the intersection between technology and business practices. First, the abilities of software-driven automation—thanks to machine learning and artificial intelligence—have accelerated dramatically with promising and frightening results alike. Second, robotic applications are leaving factory floors, opening up an entirely new world for physical automation of processes such as service robots in hotels and security robots in perimeter protection.

Again, such a development did not materialize all of a sudden; there were, like always, early signals of change. In *Robot Representation* in June 2011, Carl Telford indicated that ongoing "research could enable robots to work alongside people more effectively."[14] In *Autonomous Robots in the Wild* in April 2012, I stated, "Employment of autonomous and semiautonomous robots is becoming reality," listing a number of applications that pointed to commercialization of roaming robots.[15] Follow the strands from the future.

Here technological enablers allow companies to reshape their manufacturing and operational worlds. *Science and technology* meets *commerce and competition*, so to speak. The implications—implications that still have to play out to the fullest—affect *consumers and society*. The value-creation maelstrom at work.

Patterns Cater to Human Sensemaking

The strength of scanning stems from the robustness that patterns of events provide. A laboratory head's comments or a software developer's new application offers data points but lacks insights. A pattern of related developments guides the thinking and provides a glimpse of the future.

When similar or related developments are clustered, an interpretation of future events is not reliant on a single data point. Your insight becomes less vulnerable to misleading information. A robust pattern can put singular, even anecdotal, information in context and perspective. In other words, the precariousness of focusing on a single event is replaced by the robustness of looking at a set of related events. Patterns feature the elements you need to develop informed narratives.

Patterns can serve as important signposts on the path to future markets. As the future unfolds, participants will notice developments that fall into line with issues mentioned in previous scanning sessions; repetitive exposure is the foundation for developing awareness, creating prepared minds. As seeds of information are planted, participants will develop sensitivity to the topic. As they continue to think about the topic, they will encounter related events. Scanning can develop an itch with respect to particular topics; a sense of urgency can develop.

Scanning also can teach participants to discern weak signals by preparing the mind through frequent encounters. Infrequent exposure to new information can lead humans to miss the information. Dr. Jeremy Wolfe, head of the Visual Attention Lab in Cambridge, Massachusetts, offers the rationale behind why trained radiologists can fail to detect uncommon tumors on medical images or why airport screeners let through unusual types of weapons at security checks, for instance. He explains the challenge of rare events: "If you don't see it often, you often don't see it."[16] Scanning sharpens your senses, opens your eyes.

Eric D. Beinhocker and Sarah Kaplan, strategy specialists at McKinsey & Company, highlight the importance of creating "prepared minds." The authors explain that prepared minds are needed so that "executives have a strong grasp of the strategic context they operate in before the unpredictable but inevitable twists and turns of their business push them to make . . . critical decisions in real time."[17] Pattern identification is about preparation of mind and strategy development. It is not about predicting the future correctly. Over time, patterns bring the future into focus.

Finnish foresight researcher Mikko Dufva highlights two important aspects of scanning. He maintains that the scanning approach, or foresight, is "framing futures knowledge as a network of concepts instead of separate nuggets of information." The insight is not trivial. Nuggets of information are not patterns. Without patterns, no narratives emerge. "Knowledge as a network of concepts" will always be less vulnerable to stray bits of information; networks of concepts provide robustness. Scanning is "converting disparate 'blocks of knowledge' into perceptions of futures."[18] Such conversion elevates seemingly unrelated bits of information to a higher level of insight. Patterns redistribute the future that is already here.

Nuggets versus Crude Oil

If nuggets can be misleading, what are we looking for then? What do patterns provide us with? They make our understanding of the world more robust and detailed. They allow us to explore textures and ripples of the fabric of the future. Patterns do more. They allow us to keep tabs on the future.

Focusing on single events or exploring patterns of the future is the difference between hoping to find a golden nugget and having an ongoing operation that delivers a constant flow of crude oil. This flow of crude oil is the stream of insights you can work with. Take it; refine it into the knowledge you need.

A search for the ultimate nugget is the hope to serendipitously look at the right place at the right time. How likely is it that you will find something truly unique? I'm not saying you won't; I'm just saying, "Good luck!" In contrast, as you work with a set of forward-looking data points, you make them yours. You put them in context; you relate them to each other. You discover patterns that indeed only you might see. You certainly put them in a context only you will have.

Single events tend to stand isolated and are difficult to connect to existing operations and strategies. Patterns fit in strategic frameworks. They tell you about emerging commercial landscapes. Patterns are the foundation of narratives. Narratives enable a comprehensive understanding of market shifts and changes.

The workshops you use to form such patterns of an emerging future will extract expertise and experience from a wide range of your talent. It's where the future starts to relate to your organization's need. You make sense of implications. As participants from a wide range of domains look at the same events and developments and discuss implications for your operations, meaning materializes. Dufva highlights the relevance of such encounters: "Futures knowledge emerges from the interaction between the agents or is co-created by them." The agents are your domain experts. "New insights do not emerge from revelatory statements by visionary agents, but are rather gradually built in the interactions between agents."[19] It is not the golden nugget that unveils the future. It is not the genius at their desk. It's a team that develops patterns.

Gorbis stresses the importance of perspectives: "Thinking about the future is a collaborative and highly communal affair. It requires a diversity of views. We need to involve experts from many different domains."[20] Domain experts from various backgrounds not only inject knowledge and diversity of perspectives into the scanning process but also provide a control function. They can bring some vetting of ideas to the process at an early stage. They provide guardrails so that we don't go off course with far-flung, exciting, but ultimately impossible visions of tomorrow. Finally, diversity in people breeds diversity of perspectives. Dufva confirms "The question of expertise in foresight, in the systems view, is not only a question about specific substance matter knowledge but also about the ability to understand interconnections and change, think differently and embrace plural views."[21]

Igor Ansoff, the father of strategic management, and Edward McDonnell highlight the need not only to involve a diverse set of talent from within the organization but also to involve an even wider set of understanding. They argue, "detection of weak signals requires sensitivity, as well as expertise, on the part of the observers. This means that the detection net must be cast wide, and numerous people involved in addition to the corporate staff charged with managing issues. One source of the detectors is socio-political/economic/technological experts who are outside the firm."[22] Such a broad view is fundamental to read the tea leaves of future dynamics. (Coincidentally, Ansoff and McDonnell stress the importance of experts from fields that we found to be the areas of relevance in Figure 1.)

Diversity and multitude of perspectives are crucial. How would you be able to develop an understanding about the range of dynamics that could emerge? Also consider expanding your view even further. Outsiders to an organization can add perspectives. Intel's Andrew Grove underscores such sentiment: "The more complex the issues are, the more levels of management should be involved because people from different levels of management bring completely different points of view and expertise to the table, as well as different genetic makeups. The debate should involve people outside the company, customers and partners who not only have different areas of expertise but also have different interests." He argues, "The most important tool in identifying a particular development as a strategic inflection point is a broad and intensive debate. This debate should involve technical discussions . . . marketing discussions . . . and considerations of strategic repercussions."[23] It's the workshop where data points become issues, concerns, and opportunities. Day and Schoemaker argue, "Managers should invite employees and senior executives, as well as those outside the company who can offer relevant perspectives—such as channel partners, vendors, and industry mavens—to identify signals that may warrant a closer look."[24]

In the end, it's all about the unique insights you will be able to elicit. It is about turning a Rubik's Cube until you discover something new. It is about the picture that forces you to notice what you never expected to see. Ron Adner, professor of strategy at Dartmouth College, finds, "It is no longer enough to manage your innovation. Now you must manage your innovation ecosystem." He notes, "Expanding the perspective means changing the conversation."[25]

From Events to Implications

To make the power of pattern recognition palpable, we should look at some past changes that affected the future . . . well, now the world of today. The moment when something new emerged. The seminal developments that pushed us toward tomorrow. Like always, keep the maelstrom of value creation in mind when envisioning the future. It is here where value emerges and friction occurs.

Again—I will not get tired of stressing this part—developments in *science and technology* require careful consideration. First, relevant advances can be difficult to filter out of the constant onslaught of announcements of new insights and technologies. Such advances can be genuinely fascinating, but their true impact emerges from commercial potential: fascination cannot substitute for market relevance. Technology columnist Christopher Mims cautions, "Technology is a game of unpredictable disjunctions rather than straight-line growth or deceleration. So keep that in mind the next time a confident numerical prediction comes across your screen—just because it comes with a graph doesn't make it any more valid than any other set of assumptions."[26]

Changes in communication technologies more prominently than many other advances have changed societies and markets. For the past quarter century, the internet has transformed the way people communicate with each other. Now the Internet of Things is seeking to revolutionize communication between objects and devices.

Even revolutions start with early indications. First hints of an emerging approach to connect objects surfaced in 1999 when the Massachusetts Institute of Technology's Auto-ID Center worked actively on enabling such a network. At the time radio-frequency identification (RFID) tags attached to objects were meant to provide addressability and usability of object information for internet applications. RFID is still around such as in logistics and supply chain applications. Reduced costs of electronic components and ubiquity of wireless connectivity now enable even mere components to directly connect to the internet. Thus, RFID feels quaint today.

In June 2003, I wrote the report *URLs for Products: The Internet of Things.*[27] The report correctly foresaw the emerging connectivity of objects; it didn't get details right. RFID technology is not on the forefront of technologies considered anymore. From a conceptual perspective, it does not matter. We were able to start thinking about implications some eighteen years ago. That is not to say that we missed technological advances and didn't see changes in the emerging networks. In February 2011's *A Different Kind of "Internet of Things"* I accounted for these changes.[28] Then by May 2013, McKinsey Global Institute listed the Internet of Things as a disruptive technology and ranked the technology third among twelve technologies with the highest "Estimated potential economic impact of technologies across . . . applications in 2025."[29] In the end, we had a decade to consider implications of connected objects, components, and products. We had time to envision what such connectivity would do for services but also for security concerns.

Timeline 5 illustrates the timing of insights. In 2003, companies that were genuinely considering connectivity of objects were ahead of the curve and could start preparing for such a future. By 2013, McKinsey's report provided a wake-up call to laggards. If you didn't see it coming by then, you had work to do to catch up.

Coincidentally, in 1999, the same year in which MIT launched the Auto-ID Center that came to develop an early concept of the Internet of Things, the director general of the Office of Science and Technology in the United Kingdom discussed eScience. The former Research Councils UK in 2003 described eScience as "large-scale science carried out through distributed global collaborations enabled by networks, requiring access to very large data collections, very large-scale computing resources, and high-performance visualization."[30] It is an early description of what big data has been trying to achieve for some time now; for all intents and purposes, only the term changed, not the concept.

Again, observation of early indications of change can prepare scanning participants for a world of new concepts and dynamics, in this case for the use of massive data in a novel way. The same year Research Councils UK described the term, the report *eScience* highlighted the potential changes big data applications could bring to scientific and commercial applications. The year 2003 offered a lot to contemplate. Implications for virtually all companies were manifold.

Let's look at another example to show how patterns provide a chance to consider implications of a changing world. This example also provides a first understanding of how patterns can help build narratives. In *Robot Representation* in June 2011, Telford stated, "Players are developing a wide range of robotics technologies that will enable seamless interaction between robots and people."[31] The report provided a discussion that included telepresence robots by VGO Communications and Anybots, a Georgia Institute of Technology program to investigate how robots and humans can communicate, and various academic and government efforts to combine robotics with the at-the-time emerging concept of cloud computing. Again, a pattern provides robustness. It is difficult to figure out what the intentions of VGO's CEO are; maybe there's no commercial intent, maybe it's just a pet project of an enthusiast. But putting the information in context—developing a pattern— offers a robust look at changes. It's where weak signals of change are borne.

In September 2011, Rethink Robotics introduced Baxter, a robot for industrial settings that can sense and avoid nearby people and obstacles. My April 2012

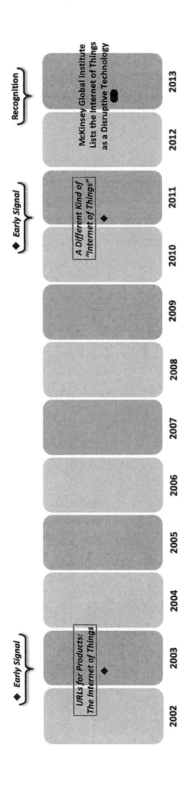

Timeline 5: Rise of the Internet of Things

report *Autonomous Robots in the Wild* then alerted to the use of robotics beyond factory settings, stating, "Cost, regulatory, and safety considerations will all play a role during the development and adoption of autonomous robots; however, such robots already see use in well-defined application areas, and their gradual proliferation has already begun."[32]

At the end of the following year, in December 2013, Knightscope was founded with the purpose of developing and deploying autonomous security robots that would monitor facilities and areas such as parking lots. The company is now known for a range of security robots that roam sensitive areas, can identify unusual noises or occurrences, and automatically alert authorities. Then in 2014, SoftBank introduced Pepper, a robot that has seen use in retail environments and can detect human emotions and respond to the feelings of its users. A cloud-based sharing system enables individual Pepper robots to tap into a large shared pool of emotional data. *Collaborative robotics*, robots that work physically inter-actively and productively with humans, became a buzzword in 2016 and now is referred to as *cobots*; early signs of such a development existed half a decade earlier (Timeline 6).

Scanning the market environment and generating stories to create early awareness of developments is an iterative process that never stops. Market developments never rest; scanning shouldn't either.

Technological advances can turn industries upside down and change entire markets. The Internet of Things, big data, and robotics certainly do. In fact, they are as much the center of hope for productivity gains—alas, advancing human culture—as they are the source of concerns, a potential beginning of a dystopian future. We can't even begin to grasp the range of implications that exist for each of the technologies, let alone how they might interact in the future. All of them lead toward automation's growing abilities.

Plus, such changes can come from far and near. Startups from one industry can develop new applications that then affect markets of previously unrelated industries. And companies' own developments can initiate the technological change that the very same companies then find difficult to adjust to—the way Kodak found it difficult to adjust to the changes of its own innovation of digital photography. Day and Schoemaker argue, "Consider that most of the technologies that will affect the business in the short run—say, within a decade or so—are in a laboratory or journal somewhere right now, perhaps even in the

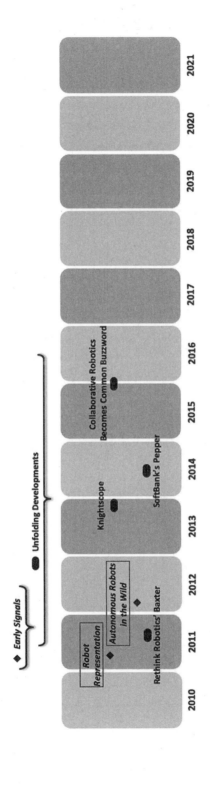

Timeline 6: Collaborative Robotics

company's own labs."[33] Indeed, consider that the future is already happening—somewhere, somehow.

But *science and technology* are one thing—enablers, as we saw previously. No one cares about technology unless they use it. And use requires more than a prototype on a scientist's workbench. *Consumers and society* make technologies into markets. "To say that technological problems cannot be approached in terms of technology alone, but must include their effects on human users and bystanders, is not to be a Luddite," Morson and Schapiro point out.[34] You want to have an impact with your technology? You need a market! And where there is a market, there is *commerce and competition*. Put them in context. Learn what the maelstrom of value creation will do.

Detecting Patterns in a Nutshell

Events and developments are data points. Patterns put them in context. Patterns offer you weak signals of change about the future. Patterns provide information.

....

Patterns make your insights tangible and robust. Beware of shiny objects. Golden nuggets of information can be misleading. Try to develop a stream of ideas, a flow of crude oil.

....

Patterns help with sensemaking. Patterns also enable communication. Patterns are valuable tools to chisel insights about the future from today's developments.

....

As you learn about dynamics and how they relate—as you develop your patterns of the future—you slowly move from mere identification of events toward understanding implications.

....

As you develop your patterns, you will be able to refine the crude oil you are extracting. Narratives will emerge that offer context. Narratives provide knowledge.

....

Keep in mind that the maelstrom of value creation requires more than only *science and technology*. You need markets, *consumers and society*. Markets come with *commerce and competition*.

....

CHAPTER SIX

Develop Narratives

EXTRACTING KNOWLEDGE
ABOUT TOMORROW'S WORLD

---■---

Imagination is more important than knowledge.
For knowledge is limited.

—ALBERT EINSTEIN

In 2004, *Conversing with Images* weaved together data about camera-enabled cell phones, internet connectivity, and consumers' interest in using images to communicate with friends and family. Four years later, in January 2008, the report *Leveraging Photography on the Web* furthered the theme. A smart observer could have seen the future falling forward like dominoes.

In March 2010, Pinterest launched a service that enables users to create image-based portfolios of interests and hobbies. A few months later, Instagram introduced its photo-sharing and posting service. Snapchat's image-sharing service went online the next year, followed by Vine's short movie-sharing service the year after that.

You can use the information you filtered to write narratives that sketch possible future alternatives. Stories bring those scenarios to life and help people remember them. As the saying goes, "Tell me a fact, and I'll learn. Tell me a truth, and I'll believe. But tell me a story, and it will live in my heart forever."

Stories also empower the imagination. It's not what you know, but what you do with what you know. Or as Albert Einstein advised, "Imagination is more important than knowledge. For knowledge is limited."

Scanning provides the elements decision-makers need to create clear, concise, and compelling stories that make sense out of what, on the surface, may seem like conflicting, discordant, and baffling events and developments.

After all, "stories have a psychological impact that graphs and equations lack. Stories are about meaning; they help explain why things could happen in a certain way. They give order and meaning to events—a crucial aspect of understanding future possibilities."[1] Peter Schwartz, who learned his trade as a futurist at the Stanford Research Institute and went on to found the Global Business Network, underscores the importance narratives have in making future worlds tangible. Narratives can convey understanding that allows decision-makers to have meaningful conversations. Narratives provide context and background.

From Pattern to Narrative

In previous chapters we looked at early patterns of big data, the Internet of Things, and robotics. We wondered how they might interact in the future. Well, all of them lead toward automation's growing abilities, which we discussed at the beginning of the last chapter. Let's use the same starting point again, except this time to develop a full-blown narrative about how the world will change through the interaction of a wide range of developments.

As new developments emerge, they won't stay in isolation. Marketplaces don't happen in vacuums. In particular, big data, the Internet of Things, and robotics are extremely complementary. Big data provides data analytics that machines can use; the Internet of Things enables remote robotics. Advanced robotics and sensor technologies feed big data applications. You get the idea.

By June 2019 it became clear that such interaction of technologies will have a tremendous impact on automation of jobs—jobs well beyond the ones mentioned in *Automation Climbing the Value Chain* only thirteen years earlier. In *Automation on Top of the Value Chain*, I looked at three developments in particular.[2]

In these thirteen years, technologies advanced tremendously. By the end of 2018 and the beginning of 2019, a completely new set of applications was in place to support—or compete with—professionals. Remember Muvee's editing software from the 2006 report? By December 2018, computer scientists from the City University of Hong Kong and Microsoft presented a deep-learning-based approach that can generate a caricature from a photograph of a person. Computers are making art now. In addition, Ahmed Elgammal, director of the Art and Artificial Intelligence Laboratory at Rutgers, is working with neural networks to create visual art. An art show in Miami, Florida, displayed some of his work in December 2018. And online art platform Artsy is featuring some of the work for enthusiasts to discover and collect art.

We've seen how Thomson Financial was able to generate simple financial reports by assembling data from news reports. Now researchers from the Massachusetts Institute of Technology and other institutions showed how a neural network can extract a research paper's core information. And this is not where it ends. The algorithm then summarizes the paper in a few sentences and also renders jargon into plain English. The result is clunky but fulfills its purpose.

Finally, in 2006, the head of the Computational Bioinformatics Laboratory at Imperial College London proposed a future in which robotic scientists will replace some human lab assistants. Well, let's see where we stand now. By 2019, robots were preparing reactions and automatically obtaining samples at the Centre for Rapid Online Analysis of Reactions at the Imperial College London. And machine learning evaluates data from the experiments.

It is easy to see how data analytics and robotics advanced rudimentary applications from 2006 into powerful automated systems. Crucially, the applications not only are better but also operate on a completely different level of abstraction and independence. *Automation on Top of the Value Chain* looks at implications. "Some of the tasks that require the highest levels of skill will eventually see automation. Traditionally secure middle-class jobs have already experienced the threat of automation, and creative, scientific, and engineering professionals will begin to see changes in their fields—some might even lose their jobs to automation."[3] We've moved beyond middle-class jobs by now. Automation's reach has expanded. An ongoing narrative has emerged, as Timeline 7 illustrates.

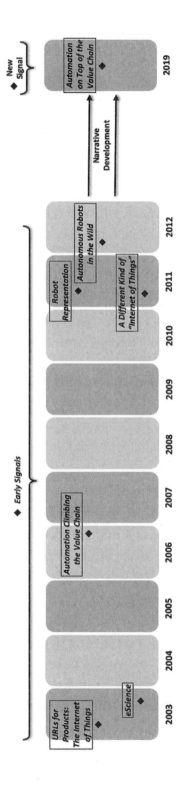

Timeline 7: The Narrative of Automation

Developments as an Avalanche

In the previous discussion, a number of technologies coalesce, creating a narrative of reinforcement. Other changes in the marketplace roll out over time and simply engulf more and more commercial space. An idea becomes a mantra; a snowball turns into an avalanche. Timeline 8 illustrates how a concept can gain steam over time and become an avalanche that is engulfing more and more use cases.

Snowballs can start with individuals. In 2011, the quantified self movement held its first, at the time still very small, international conference in Mountain View, California. Followers of the movement promote the capture of physiological data for personal analysis. The same year, *Self-Tracking Health Data* found, "The Quantified Self . . . has gathered a sizable following."[4] The report pointed to initial commercial business opportunities of the approach. The movement ultimately led to an entire product and service market of fitness trackers and related apps. The future of wearables is still shaped by the idea of a quantified self.

Over time, the concept lost some of its quirkiness. Serious use became a consideration. Quantified self moved toward *diagnosed self*—the use of fitness, health, and emotional information to make inferences about health issues. My report with the same title from 2014 highlights how practitioners can "benefit from making more informed decisions about their health, identifying potential health issues early, and creating data records that medical practitioners can analyze." It also warned of "making inappropriate self-diagnoses, experiencing constant paranoia about potential disease indicators, and offering ill-informed treatment advice in online forums."[5] Benefits and concerns have not gone away since.

Once the seed was planted, the concept rolled over applications and markets. The concept moved from individuals to enterprises to community groups. My report *Quantified Everybody* from 2014 listed developing privacy concerns.[6] In January 2016's *Quantified Employees* I looked at enterprise uses of the concept.[7] Previously, in 2014, *Quantified Communities* discussed how the "design of future urban infrastructure depends not only on civil engineers but also on data scientists, environmental engineers, and smart-technology professionals."[8] At the time, real estate developers of Hudson Yards, a neighborhood in New

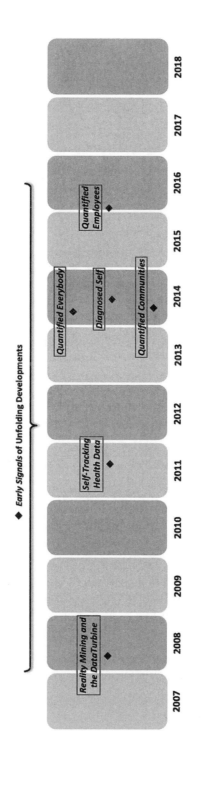

Timeline 8: Quantifying Individual and Societal Behavior

◆ Early Signals of Unfolding Developments

Reality Mining and the DataTurbine

Self-Tracking Health Data

Quantified Everybody

Diagnosed Self

Quantified Communities

Quantified Employees

2007 2008 2009 2010 2011 2012 2013 2014 2015 2016 2017 2018

York, New York, collaborated with New York University's Center for Urban Science and Progress to measure and model a number of factors, including air quality, energy use, inhabitants' health, pedestrian movements, recycling, traffic flow, and waste disposal. The director of the university's center, Steven E. Koonin, claims the project will support data scientists in developing new ways to model communities. (By 2017, Andorra—a small country of 77,000 people between France and Spain—started serving as a "living lab" for researchers from the MIT Media Lab's City Science Initiative to test urban models in a real-world context.[9])

The concept of capturing an increasing amount of data on human behavior and environments didn't come from nowhere—nothing does. Early alerts are there for you to uncover. In 2008 Alex "Sandy" Pentland, a professor at the Massachusetts Institute of Technology, presented the concept of "reality mining."[10] He looked at ways future data applications could extract valuable commercial data. The same year in June, *Reality Mining and the DataTurbine* not only outlined the concept but also pointed to commercial implications as the title implies.[11]

Developments as a Vortex

Other narratives look at commercial developments that resemble a vortex that gobbles up more and more industries. Let's take a look at the sharing economy. The concept has morphed and changed over time. What was once an idea driven by individuals who wanted to avoid monetary exchanges now is big business and a controversial money machine. Times change.

The *sharing economy*—roughly defined as consumers' sharing of resources they own—has had an impact on the marketplace, the startup community, the fabric of venture-capital spending, and incumbent companies and resulted in a growing cottage industry of sharing. The sharing economy initially emerged from consumers, was created by and among consumers, and focused on the needs of consumers. Only over time did entrepreneurs and investors pick up on this development. The development of sharing networks and their providers' ascent to global competitive forces did not come out of nowhere.

Companies like NeighborGoods, Share Some Sugar (which no longer exists), and Snapgoods (which changed its business model and service offer) enabled neighbors and friends in a network to share tools and backyard equipment, among other things.

Early alerts existed—again. In *Next-Generation Consumers* in July 2009, Brock Hinzmann saw, "A redefined understanding of purchasing, owning, and sharing could turn out to be . . . persistent."[12] The same year in August, in *Community Values Trump Individualism*, Aster Peng argued, "A new, profitable target may be the market for family, community, or otherwise shared products and services."[13] Then, in December of the following year, in *Collaborative Consumption*, Kimberly Wiesbrock found, "Some consumer segments are developing a new understanding of ownership and are redefining consumption, moving toward a more collaborative approach to product use."[14]

No doubt, something was brewing—long before sharing-economy players Airbnb, Lyft, or Uber Technologies became household names, in fact before two of them even existed. Airbnb and Lyft started operations in August 2008 and July 2012, respectively; Uber Technologies went live in July 2010, although the company had conducted prior work. In 2013, CeBIT in Germany, the largest and most international trade show for computer-related products, chose as its motto "The Shareconomy." Arguably, this is the point when everybody's awareness gap closed; sharing was now common knowledge.

By the end of the 2010s, sharing had become an economic driver rather than a communal activity. By the end of 2019, Airbnb's market capitalization was estimated at well over $30 billion. And at the beginning of 2020, Lyft and Uber Technologies' market capitalization were at about $15 billion and an astonishing $70 billion, respectively. Or over $100 billion for the three companies. (Then Airbnb went public at the end of 2020 and hit almost $100 billion alone within days.)

Although many observers tend to look at these companies as technology providers, they are not. An app doesn't make a high-tech company. They are intermediaries. They use technology to leverage emerging consumer behavior— the value creation maelstrom at work. This development is shown in Timeline 9.

Some observers saw indications even before my team reported on the development. In June 2005, Robert D. Hof, Silicon Valley bureau chief for *BusinessWeek*, wrote in "The Power of Us" about companies that started to use communities of consumers for crowdsourcing efforts. Hof also presented others' prescient

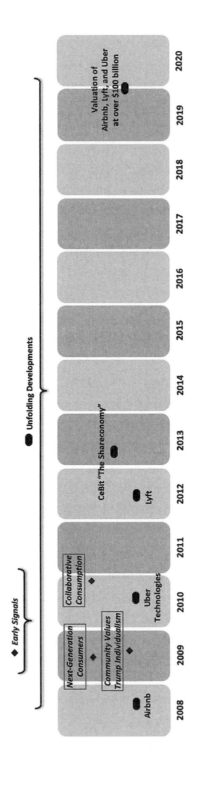

Timeline 9: Sharing Economy's Ascent

observations. C. K. Prahalad, professor at the University of Michigan, commented, "We are seeing the emergence of an economy of the people, by the people, for the people." And Yochai Benkler, professor at Yale Law School, voiced a new mode of production that is emerging and goes beyond the traditional firm-market model, calling it "peer production." Finally, Tim O'Reilly, a publisher of technology books, mentioned the "architecture of participation."[15]

I believe that a full-blown understanding of concept and reach of the power of us evaded even these experts at the time. But a move toward individuals' economic participation was palpable. And in September 2005, I wrote in *From Cocreation to Competition*, "Customer cocreation is the latest rage, but companies will need to develop ways of accommodating cocreation models without suffering marginalization."[16] I do believe now—looking back—that marginalization perhaps was a strong word, but a word that certainly strikes a chord with some companies that have become increasingly affected by the sharing economy.

Developments as a Gravitational Force

Finally, some concepts simply replace old ways of doing things. Moreover, they develop a gravitational force that attracts use from all over the place. They gobble up applications. They change consumer behavior for good. In the process, industries change, morph, or vanish.

Although predecessors to Apple's smartphone existed (and foreshadowed the era of smartphones), it was the introduction of the iPhone in June 2007 that changed consumers' expectations. A multitouch user interface, multiple sensors, and a convenient way to integrate novel apps on the fly via an app store grabbed users' attention.

Smartphones have become our tricorders, advanced tools as envisioned by the *Star Trek* series half a century ago. In the process of advancing from telephone to tool, industries were surprised and businesses ambushed. Did such ambushes happen all of a sudden? Or could such a development have been anticipated?

In *Unexpected Competition from Smartphones* in March 2009, I highlighted impending changes because of mobile phones' newfound utilities: "By becoming more advanced, smartphones are rapidly moving beyond the telecommunications market and competing in a wide range of markets against a host of products and services, most of which only five years ago would not have considered themselves potential victims of cell phones' newfound capabilities."

In particular, I highlighted a list of product categories from distinct industries: "The market appears to be gravitating toward smartphones, which means that portable game platforms, eBook readers, MP3 players, and portable navigation devices now have to face a formidable new competitor."[17] The report also mentioned dedicated camera sales, which were already expected to decline at the time. In short, the report called attention to smartphones' ability to eliminate—or at least drastically reduce—the market for dedicated devices.

Headlines and discussions in succeeding years trace the report's anticipated pathway that smartphones would take. In January 2014, Hollister's "The Age of the iPod Is Over" detailed the MP3 players' demise because of smartphone use as music players.[18] In October 2015, an editor's (Paul's) experience when using navigational devices on a road trip culminated in the headline "Traditional GPS Is Dead. Long Live Smartphone GPS."[19] In March 2017 in "This Latest Camera Sales Chart Shows the Compact Camera Near Death," Zhang stated, "As smartphone cameras continue to improve, compact camera sales continue to nosedive."[20] Then, in January 2018, in "The E-Reader Device Is Dying a Rapid Death," Haines proclaimed, "Smartphones and tablets are the choices of ebook readers."[21] Finally, Wijman's "Mobile Revenues Account for More than 50% of the Global Games Market" highlighted the success of mobile games on tablets and smartphones with phones accounting for 51% of the entire gaming market for 2018, not even listing dedicated mobile gaming platforms anymore.[22]

All of a sudden? Not really. There were indications that such developments might occur. Some patterns are narratives all by themselves. In 2009, observant decision-makers could have been ahead of the curve. A decade later, the smartphone's gravitational power was old news, as Timeline 10 shows.

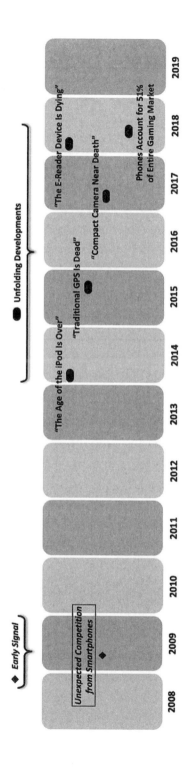

Timeline 10: Smartphone's Competitive Impact

Early Signal

Unexpected Competition from Smartphones

Unfolding Developments

"The Age of the iPod Is Over"

"Traditional GPS Is Dead"

"Compact Camera Near Death"

"The E-Reader Device Is Dying"

Phones Account for 51% of Entire Gaming Market

2008 2009 2010 2011 2012 2013 2014 2015 2016 2017 2018 2019

How Did We Get Here?

Let's recap. We move from events and developments to patterns to narratives. Our journey takes us from data points to information to knowledge. We learn to understand what the future is made of.

Judiciously selected—filtered—events and developments enable us to glimpse at future commercial dynamics. It's the information we work with. This information keeps us grounded in reality. Using it keeps us from making up futuristic imaginations. We don't have to fabricate dynamics. Existing events and developments make you find an anchor point to attach your narratives to. Moreover, if I ask you why you came to a set of implications, why you reached certain conclusions, you can always show me the evidence you started with. You prove credibility. The events lay the foundation the future will be built on.

Now good data points come with certain qualities. They are accurate: yes. They are relevant: true too. But they provide us with anchor points. Anchor points that allow us to connect the dots of tomorrow's fabric's strands. These anchor points hold our understanding together. In fact, good data points provide us with multiple anchor points to think about technology, competition, and society all at once. They allow us to weave stories from the future.

Patterns outline such connections. Where do events overlap? How do developments relate? Connections provide meaningful implications. Such connections provide understanding. We learn how the world of tomorrow might come together. We envision braces and brackets that hold up tomorrow's house.

Finally, narratives tell us a story. They allow us to visualize tomorrow. They enable us to communicate the future. Narratives let us anticipate the structure and architecture of the house in which commerce will take place in the future.

Visualizing a wide range of dynamics that could shape tomorrow is a necessary requirement to make forward-looking decisions. It is not sufficient though. You have to consider implications for your corporation—after all: one person's threat can be another person's opportunity. And you have to let actions follow—obviously, one hopes.

Narratives Build Worlds

Forecasts limit your view. Big data looks at past information. Uncertainty does not afford a single perspective on the future. What is needed are worlds that allow you to roam freely to contemplate your options.

Ralston and Wilson provide the rationale: "While traditional responses—improved forecasting capability, better monitoring, greater flexibility—are needed and commendable, they are inadequate to meet the full extent of this challenge. What is needed is a system geared to speculate not only about 'the future,' but about a range of possible futures."[23] What is needed is FIPI that lets you look at developing patterns and narratives to draw implications across the wide range of possibilities that the future offers from today's point of view.

Dufva looks at foresight, including scanning, "as a *systematic* practice for exploring futures: foresight is a process where a set of methods is used in a planned and rigorous manner (systematically) in order to create futures knowledge."[24] Crucially, futures knowledge does not represent "deterministic forecasts." Futures knowledge is a set of "justifiable contingent plausibilities." You want justifiable narratives to make sure that you are credible. Ground your thinking in realities; let real events and developments be your starting point in FIPI. Make narratives contingent; present dynamics to tell a story that captures decision-makers' imagination. Finally, do not fall into the trap to tell stories that veer off into worlds that listeners cannot follow. At the same time, don't attach likelihoods. "Make It Plausible, Not Probable"—Oxford University scholars Angela Wilkinson and Roland Kupers highlight the mantra of anticipating futures.[25] Whereas probability creates expectations, plausibility builds preparedness.

Traditional forecasts prevent comprehensive decision making. They present a very narrow look at the future. They limit your choice by showing you a narrow pathway. They lead you toward dangerous territory by excluding uncertainty. Narratives of potential futures, in contrast, open the corporate mind to vast opportunities and a virtually unlimited number of strategic alternatives that could find use. The challenge is obvious. You will see very different worlds that will require different responses to stay ahead of the game. But such narratives reflect reality; they present the future as it is—uncertain. But looking at alternative futures allows you to develop organizational intelligence. After all—according to American author F. Scott Fitzgerald—"the test of a first-rate intelligence is the ability to hold two opposed ideas in the mind at the same time, and still retain the ability to function." Scanning is the support to retain the ability to function.

Narratives Suit Human Understanding

Narratives of marketplace dynamics tend to be more holistic and systemic if diverse teams process data and discuss implications. The more perspectives find a voice, the more connections and relationships between developments can be uncovered. Morson and Schapiro introduce the concept of "narrativeness, which comes in degrees, [and] measures the need for narrative." The concept enhances what narratives can achieve. "The more we need culture as a means of explanation, the more narrativeness. The more we invoke irreducibly individual human psychology, the more narrativeness. And the more contingent factors—events that are unpredictable from within one's disciplinary framework—play a role, the more narrativeness."[26]

The term is unusual, but its meaning is not. People make markets. Contingencies are what create uncertainty. Narrativeness describes what good stories should look like. Good stories create worlds we can accept as a plausible future. They represent a scenario of tomorrow. They don't predict tomorrow but tell us instead what we should be looking for and be concerned about.

In an increasingly connected and complex world, narratives provide an intuitive way to convey knowledge and insights. Scanning focuses on making connections between developments, finding relationships between events, and identifying areas of friction in the market. Many corporate decision-makers zero in on predictions and forecasts but don't realize that such a focus only limits your view on the future. Narratives, in contrast, allow decision-makers to explore dynamics and to spot opportunities that forecast-focused competitors will miss. The Joint Research Centre of the European Commission notes a shift in futures studies more generally "from emphasis on predictive approaches to more exploratory studies, and from one-off studies to more continual iterations of the process of envisioning future challenges and opportunities."[27]

The importance of human understanding is a crucial and enabling aspect of the scanning concept. At its core, scanning is about inferences from events and dynamics through the lens of people as market protagonists. Madsbjerg cautions, "Never before has our world of overlapping political, financial, social, technical, and environmental systems been so inextricably linked. We must remind ourselves—and the culture at large—why the human factor is the most important factor when it comes to making sense of this world." The human element is not the weak point in scanning that requires elimination; it's the crucial ingredient that enables scanning to peek into the future. Madsbjerg explains, "If we don't have a perspective on the human behavior involved, our insights have no power."[28]

Stories also have a strength that other sources of information miss. They are easy to communicate and to understand. An evening at a campfire comes to mind; it's second nature for most of us. Bernardo Blum, Avi Goldfarb, and Mara Lederman, at University of Toronto's Rotman School of Management, highlight the human component: "To extract the full value of their data investments, organizations need to build tighter links between data analytics and everyday business conversations."[29]

According to the *Handbook of Knowledge Society Foresight*, "Some people find offputting more than the most basic statistical information; many find it difficult to examine the data critically (or at least in an informed way)." It continues, "The most sophisticated quantitative methods require considerable expertise to apply [and] many assumptions about the nature of the data and the most appropriate methods of analysis are concealed in statistical techniques, and it is common for data analysts to follow common practices rather than to examine whether they are really adequate for the task at hand."[30] Narratives are a cure to such illness.

Narratives Highlight Change

Here is the crux of our efforts. Here's why we want to go through all the steps to build our stories and narratives: extrapolations "fail to assess underlying driving forces, so that there is an inability to anticipate changes in these forces." Also, "they do not examine whether qualitative transformations might disrupt, or radically modify the meaning of, change in quantitative indicators."[31] You want to find disruptions, look at the stories small data can tell you. Schwartz reminds us, "Stories . . . give order and meaning to events."[32]

Narratives can convey understanding that allows decision-makers to have meaningful conversations. Narratives provide context and background. Narratives supply decision-makers with the required knowledge to take necessary actions.

In Amazon's 2017 Letter to Shareholders from April 18, 2018, Jeff Bezos provides a separate section, "Six-Page Narratives," to highlight the importance of narrative. "We write narratively structured six-page memos. We silently read one at the beginning of each meeting in a kind of 'study hall.' Not surprisingly, the quality of these memos varies widely. Some have the clarity of angels singing. They are brilliant and thoughtful and set up the meeting for high-quality discussion."[33]

An open mind and willingness to take a broad look are crucial when anticipating the future. Scanning workshops importantly enable participants to probe potential futures. Experts' interactions can scratch the surface of marketplaces' veneers. What is happening underneath is what you need to know.

The process up to now is as powerful as it is intuitive. Patterns and narratives cater to human sensemaking. Humans, simply because of their need to navigate the world from birth on and because of—not despite—their limitations, have learned to filter information efficiently, identify patterns effectively, and surface connections and dynamics intuitively. It's how we work.

Madsbjerg outlines the rationale of such sensemaking: "We can think of sensemaking as the exact opposite of algorithmic thinking: it is entirely situated in the concrete, while algorithmic thinking exists in a no-man's-land of information stripped of its specificity. Algorithmic thinking can go wide—processing trillions of terabytes of data per second—but only sensemaking can go deep." He continues, "With sensemaking, we use human intelligence to develop a sensitivity toward meaningful differences—what matters to other people as well as to ourselves."[34]

To see the next disruption coming, we need specificity. To go beyond the obvious, we need to go deep. Narratives matter. They are not sufficient though, as we'll see in the next chapters.

Developing Narratives in a Nutshell

Narratives package information in accessible stories. Narratives leave data points behind and create knowledge.

....

Narratives are easy to communicate. Narratives enable sharing of crucial dynamics in today's world. Narratives can provide a powerful sketch of the future.

....

Scanning guides you through the process of developing such narratives. Only a handful of crucial pieces of information—small data—give you the story of change—the alert of big disruptions.

....

Big data and algorithms remove markets from decision-makers' minds. Narratives allow sensemaking; we can see the cogs and wheels at work. We can sense the unfolding of change.

....

Powerful stories start with today's developments; they ground us in reality. Stories explore how these developments connect. Stories tell us how they could roll out over coming years.

....

Narratives provide awareness; they are necessary to make us understand what is in store for tomorrow. We still need to identify implications and the actions that will prepare us for tomorrow; the work is now beginning.

....

CHAPTER SEVEN

Identify Meaning

MAKING TOMORROW'S WORLD RELEVANT

———————————————■———————————————

If you do not expect the unexpected, you will not find it.

—Heraclitus of Ephesus

You're scanning the business media for events that might affect your market strategy and quickly dismiss some little blips that will not alter the future dramatically: marginal improvements in computer chip performance (even though continuous advances over more than half a century transformed the world), a slight swing in an appliance manufacturer's market share, and the fluctuating price of commodities over the years.

Then you learn about a group of scientists who are using machine learning to discover new chemical compounds, while another group is pioneering a data-driven facility to automate research into chemical processes. Interesting. You discover that scientists are using machine learning to predict magnetic properties of novel materials and that a government institution is using AI to automate molecule discovery. Suddenly, you have wrangled all the ingredients for a strong narrative about the astonishing ways in which materials science might usher in a whole new era of automated exploration and discovery. Now these ingredients have gone from interesting to relevant. No matter what your business (think mining, logistics, design, manufacturing, telecommunications, and health care), this development will affect you and your business.

While FIPI's filtering and identifying steps apply to every company across all industries, the prioritizing and initiating steps will vary from company to company. In the case of automated material development, an aerospace and an electronics company will respond to this game-changing event in their own unique ways.

You need to develop implications—implications for you and your eyes only. Scanning offers an intuitive way to do that: the impact/emergence matrix, which is an easy-to-use tool that enables decision-makers to sort narratives and to visualize their relevance to their organization. It helps define the degree to which the development might affect your organization and the likelihood that it will turn into a major game-changer.

An Arm's Length Away Is in the Center of the Action

In 1998 *Electronic Delivery and Distribution of Music* highlighted a new way to send music files. More important, the report showed how such a change in logistics would ripple through the industry. After all, arguably, the one ace the industry had was its access to the market. Now a new technology enabled an alternative approach. Well, actually a superior approach, although transmission speeds were still at a snail's pace at the end of the last century. But the crack had occurred—the crack that would lead to a full-blown disruption. After all, who wants to go to a record store if you can just click a link on your computer in your living room? The music industry was stunned . . . and it was only the first industry that would be surprised.

Other media industries initially looked at developments in the music industry as some sort of aberration, perhaps even with some kind of amusement. After all, book publishing was a different animal. Who wants to have a clunky computer on their lap instead of nicely textured paper product? And then came ereaders and tablets. And how could movie files ever be transmitted over the internet if a menial song takes minutes to arrive? And then transmission speeds exploded. You might still consider yourself safe although you're only an arm's length away. Always consider: an avalanche starts somewhere far at the top and inevitably you will be rolled over at the bottom. A vortex spirals slowly from the center out, and gravity has reach; an arm's length does not matter much.

The publishing industry slowly saw the arrival of ebooks, digital copies that can instantly ship and are readily available on ecommerce sites. Libraries introduced temporary rentals, some kind of streaming if you will. Google's (whose domain name was registered in 1997) introduction of its search engine—not the first, but arguably the most powerful—unfolded its effect on information providers, library services, and clipping services with might. Wikipedia, launched in 2001, single-handedly reduced the market for encyclopedia providers to rubble; the last printed version of the *Encyclopædia Britannica* was published in 2010. Online publications of text-based media continue to put pressure on newspapers, magazines, and journals. Some argue the effects on journalism in general are game changing. Think about how you got your reading material in the 1990s . . . if you were around then.

The movie industry thought file sizes of visual media would protect them. And then Blockbuster and Hollywood Video were hit hard only a decade after the music industry felt the pain. Netflix moved to streaming services, turning the industry upside down. Hulu provides a competing service, and YouTube is another alternative. Now Apple, AT&T, Disney, ESPN, and others all see the value of streaming. Plus, Apple and Netflix have started spending billions on their own content creation. What a change!

The video gaming industry moved—more successfully than other media types—to completely new distribution, customer-interaction, and revenue-generation models. Many video games in fact have moved to the internet. How easily can you connect with players from all over the world?

Timelines were different for the various industries. Competitive considerations varied. Crucial aspects are the same. New competitors changed up the playing field. Consumer expectations changed rapidly. Pricing strategies and revenue models fell apart. Piracy concerns and copyright worries became business-model-threatening issues. There you have it: disruption at its fullest.

All mentioned industries—and many adjacent ones—were affected by file sharing and streaming. Did these technologies affect all industries to the same degree? No, some businesses are gone now, others cling on by the skin of their teeth, and a whole bunch—of mainly new entrants—have managed to thrive in the new environment. Did all the industries face challenges at the same time? Again, no. Technology had to advance; consumer behavior had to catch up. The new way of doing things took a good decade to ripple through industries.

It seems impact and timing could be good tools to assess disruption's effect on an industry. Such criteria work even better for your organization. You know your strengths, weaknesses, and challenges best.

It's Your Turn

An industry tends to share many challenges and fortunes. But individual companies remain unique. Your company is unique. So, don't fall into the trap of associating yourself too closely with the industry you consider yourself currently to be a part of. You have unique strengths; you have weaknesses others might not have. Implications of disruptions are different for you than they are for your competitors. Let's see how we can judge what a disruption will mean for you.

When you're exploring the dynamics and developments in the marketplace, when discovering what the future might bring, focusing back on today's decisions can be difficult. You became aware of changes and disruptions that might occur over coming years. But what does that mean to your organization? What should your company's goal be in a changing world? What do you need to consider when making your next moves?

So, here we are. We're now moving toward the third step: prioritizing issues you identified by looking at the implications for your organization (Figure 8). This step is also the one that moves us away from our focus on developments and events we have no influence over. We're now moving away from the commercial environment we will have to operate in toward what really matters: the decisions and actions we will have to take.

No organization can afford the luxury to focus on every blip in the market fabric simultaneously. Your organization needs to make a judicious decision regarding what developments are worthwhile considering. It's about relevance of developments; it's about the priorities of topics. It's about putting your precious time, effort, and resources where they really matter.

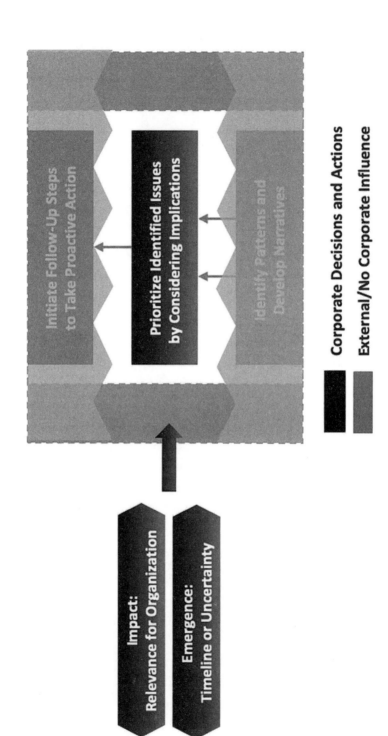

Figure 8: Step 3 Highlights Critical Issues

The process requires the identification of topics most relevant to an organization as well as a hierarchy of priorities among these topics to focus on. *Filtering data points* and *identifying patterns and developing narratives* look at the futures you might operate in. The following steps—*prioritizing issues* and *taking actions*—are about you. Now that you're aware of changes and disruptions, how do you prepare your organization? It's where the proverbial rubber meets the road.

Understand What Matters

Assessing the relevance of issues that will affect an organization is a common task. To do so, we can borrow a handy tool from the broader world of strategy development. It's understood, it's intuitive, and it very naturally moves us toward taking actions.

Two aspects deserve consideration: impact and emergence (perceived uncertainty or timeline) of the development. They form a matrix with impact as the vertical y axis and emergence as the horizontal x axis (Figure 9). Impactful topics move to the top to catch attention. They show us what deserves our attention. Sorting topics by emergence will help us understand what next steps should be. This task shows us how we should deal with challenges.

There's no magic to the matrix. Its straightforward design makes it easy for everybody to get involved fast to make sense of the world. And, as we will see later, it allows communicating results as a shortcut to the complexities of the future. It's a sketch from the future.

Impact is a fungible term that can accommodate a variety of parameters and metrics. For instance, judging impact according to profit margins as opposed to market share can lead to vastly different assessments. Definitions of impact can include sales volume in units, or revenue, or its growth rate. The matrix can also accommodate softer criteria—although a clear definition might be challenging; be careful with misunderstandings. For instance, security considerations, regulatory challenges, and brand development can all be criteria. If you wish to apply a much broader definition—such as general strategic importance, as most companies tend to do—you are free to do so. Be aware though that your team members might have a very different understanding of what that means; comparing apples

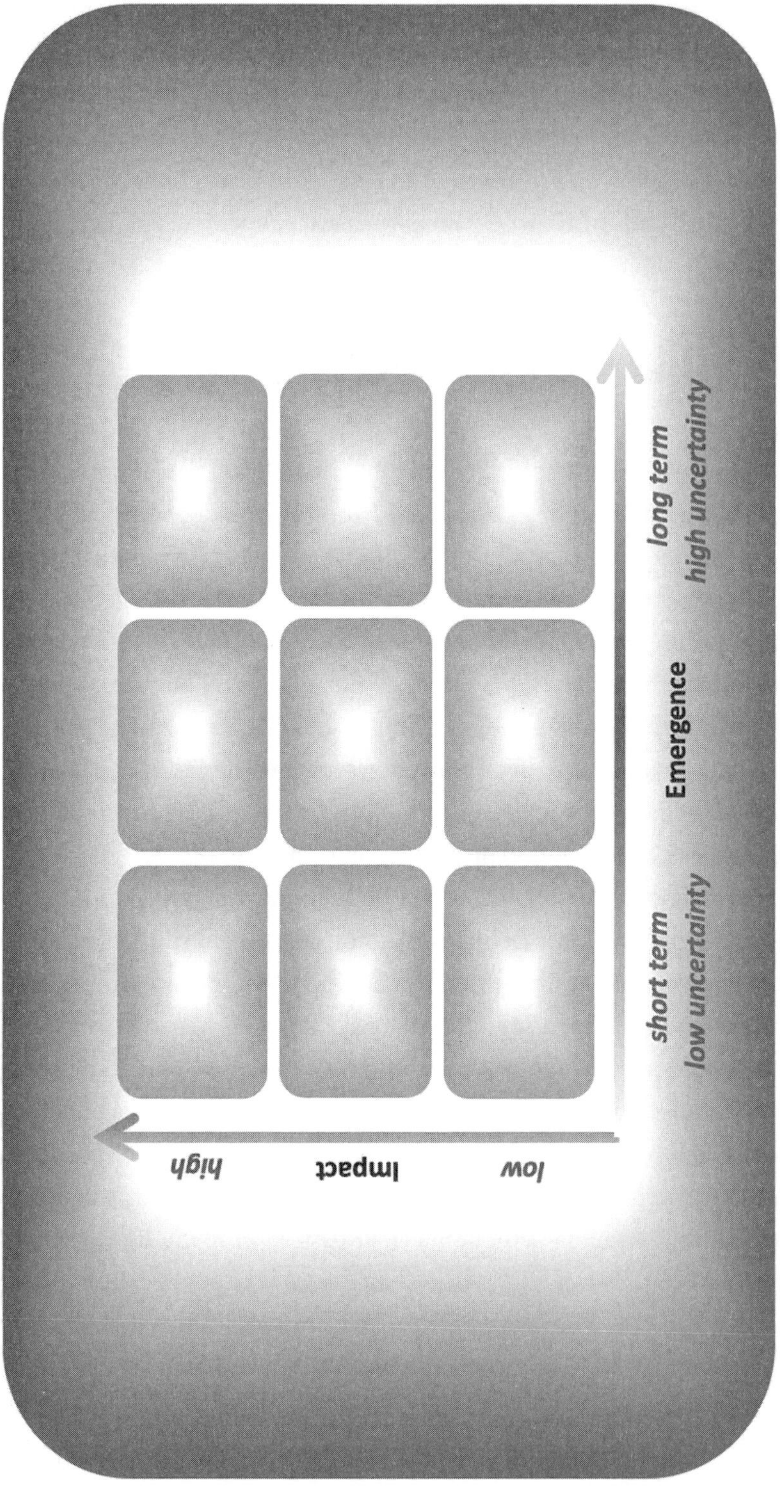

Figure 9: Impact/Emergence Matrix

to oranges is a danger. The matrix will help you get everybody on the same page; that doesn't mean that everybody is on the same page going in. Whatever your choice, keep an eye on your particular corporation's situation.

Emergence can address two different considerations: the level of uncertainty you perceive and the expected timeline on which the development will unfold.

Uncertainties can feature very different flavors. Some uncertainties can be captured in an either/or category: either a development will happen or not. Other uncertainties relate to a range of outcomes. And then there are situations in which a very wide range of outcomes is conceivable; sociopolitical developments of countries and regions usually are of that type. Uncertainty is low when the participants generally share a comparable view about future developments. In contrast, uncertainty is high if participants are generally unclear or have conflicting opinions about how the issue will develop. Crucially, "uncertainty is not simply the absence of knowledge. Uncertainty can still prevail in situations where a great deal of information and knowledge are available."[1]

Some participants will find the concept of uncertainty difficult to judge and will prefer to refer to a timeline for the short term, perhaps already occurring, to the long term, depending on the industry at some time five to ten years in the future. A technical difference exists between uncertainty and timeline, but you shouldn't get bogged down by attempting to be scientifically accurate. You won't be able to achieve that accuracy here; later research can provide you with details. Here it's tomayto, tomahto; you will find that both criteria will lead to an outcome that's agreeable—not only agreeable, but very intuitive.

Timelines of developments and the degree of advance notice a company requires vary substantially across industries. In software development and smartphone manufacturing, half a year to a year of advance notice constitutes victory. These industries move fast; changes are implemented rapidly. Then let's look at mining operations, which have a completely different time horizon. Investments pay off over decades. Timelines and uncertainties play out very differently for these industries.

And Now for Something Completely Different

Some topics could throw you off. With good reason, books have been written on them. A selected few topics feature an intriguing conundrum; they are highly uncertain and extremely speculative but would have a tremendous impact on everything that follows. They are wild cards that deserve separate consideration. Emerging quantum computers fall in this category. They would change how we can capture and calculate the world. But they also would eliminate many security mechanisms we designed for the internet. Imagine every transaction on the internet could be breached, hacked, or changed.

Wild cards are events that are difficult to deal with. Some define wild cards only as developments that occur suddenly with instant large impact on markets; others include developments that unfold gradually but develop tremendous impact similar to a tanker that runs into a pier. I am not suggesting that there truly is a right or wrong way such concepts are used; I'm just saying it's all the same to me. We're talking events that change the fabric of marketplaces—sometimes even humanity—as we know them. Be aware that they exist; be aware also that preparing for them is a completely different ballgame, often impossible.

Whatever your definition, the strength of the effect on the marketplace counts; the degree of the change in dynamics matters. Wild cards represent extreme strategic challenges, no matter the time required. If they occur suddenly, the element of surprise can leave decision-makers blindsided; if they unfold their market-changing power slowly over time, decision-makers will find it difficult to identify their true potential for impact before it is too late.

Wild cards are signals that can unfold their relevance in a single event, which is the result of preceding developments. The explosion of the first nuclear bomb suddenly and dramatically threw the world into the nuclear age for better or for worse. In contrast, the expectations of quantum computing are growing over time as researchers start to develop a better understanding. Nevertheless, the first time quantum computing will provide usable results beyond mere scientific experimentation, its genuine impact on markets and applications will be difficult to fathom from today's point of view.

Wild cards tend to be singular events that have the power to affect the future by themselves, whereas less impactful events develop relevance only in conjunction with additional events that then together accumulate market relevance. That does not mean you won't be able to anticipate them; it just makes prioritizing them in a meaningful way or finding appropriate actions next to impossible. Likely, your decision making will start to spin.

In the case of 9/11 and 3/11, surprise, impact, rapidity, and scope were all present. When the South Tower of the World Trade Center collapsed at 9:59 a.m. on September 11, 2001, the world became instantly aware that something significant had occurred. A particular minute pinpointed the moment the world changed. The scope of the changes was tremendous. Geopolitics, legislation, military actions, security, safety, law enforcement, and surveillance activities—only for beginners—never were the same.

Similarly, the Fukushima Daiichi nuclear disaster on March 11, 2011, had a very rapid development. Within not even a day of the earthquake that caused the tsunami that damaged the nuclear reactor, the core of reactor 1 had melted, releasing radioactivity. Again, the world learned very quickly that substantial changes would occur. The disaster had a massive impact on nuclear policies globally. The level of surprise and scope of the events, however, was mitigated by previous experience with the Chernobyl accident in 1986. (Then again, a previous World Trade Center attack on February 26, 1993, with a truck bomb was a sign that the structures were terrorist targets.)

In *Predictable Surprises: The Disasters You Should Have Seen Coming, and How to Prevent Them*, Bazerman and Watkins make a convincing point that 9/11 should not have been the unexpected surprise it turned out to be.[2] And as early as 1990, the US Nuclear Regulatory Commission warned of cooling system failures in seismically active regions—a report the Japanese Nuclear and Industrial Safety Agency cited in 2004. And even the Manhattan Project that culminated in an instant explosion to global awareness took years to develop. (Also, the US wasn't the only country to consider such a bomb.) Even wild cards cast their shadows beforehand.

The question though is what to do with awareness of such possibilities. Translating awareness into preparedness is the crucial strength of FIPI. Unfortunately, wild cards make sensible positioning on the impact/emergence matrix nonsensical. If it happens, the impact will be all-encompassing. But when it happens—if it happens—is unclear. Do you want to change all your operations for a distant plausibility? Do you want to put all your money on one lottery ticket? I didn't think so.

Taking wild cards into account is challenging. The moment at which decision-makers have to respond to changes is difficult to pinpoint. Here's an extreme example: a sudden meteorite strike on London, New York, or Tokyo would change global commerce as we know it. The effect would be felt globally—physically and economically. It's a wild card, all right. But how would you prepare for it? Quick responses and ongoing reactive navigating of constant changes in the fabric of changing markets likely would trump any strategic plans you could imagine. Sometimes reflexes count more than meticulous contingency plans.

In contrast, the internet was in development for decades if you consider all the strands of communication technologies and computing as early signs. Even at the time the first message was sent on October 29, 1969, no immediate commercial effect was unleashed. The impact was unleashed over time. In truth, the question of when decision-makers had to take the internet's capabilities into account depended on the industry and particular company.

Then there's mere credibility of process within your organization. It's easy for your team to question the entire exercise when considering the unmanageable. Saffo underscores a methodological challenge of considering such events: "The tricky part about wild cards is that it is difficult to acknowledge sufficiently outlandish possibilities without losing your audience."[3] This juxtaposition creates a conflict. On the one hand, scanning is trying to include a wide range of plausible events, but on the other hand, and perhaps even more importantly, process and related activities have to find acceptance, ideally enthusiasm, within an organization. The solution lies in pragmatism of approach and execution.

False Wild Cards

At the beginning of 2020, the spread of the coronavirus slowed down the world economy—getting it to a previously inconceivable standstill. Does such a massive impact of a single event—the pandemic—automatically make for a wild card? Many commentators thought so. Some called it a black swan, others a wild card. Many observers argued that the spread of the coronavirus was a perfect example of an event that was unpredictable (because of a perceived low probability, I assume). They also mentioned the unexpected impact on the global community. But neither point in this argument is accurate, and the distinction of whether we are talking about a wild card or a foreseeable event—or predictable surprise—matters a great deal from decision-making and risk-management perspectives.

Here's the deal. The occurrence of such a virulent virus was not only expectable but also predictable in the sense that the emergence of a virus-related health crisis should be unsurprising (details such as type, timing, and starting point of the outbreak clearly are beyond predictability). It's like asking, "Did you foresee the earthquake in California?" Well, I can tell you that it will happen, so you'd better be prepared.

In the case of the pandemic, there's a well-documented precedent. The Spanish flu killed from 50 million people to as many as 100 million people about a century ago (at a time when the world population hadn't even reached 2 billion people). That example also should put the argument of unexpected impact to rest. Too far in the past? Then let's look at this century, the past two decades, and try to find the breadcrumbs that should have alerted us. In fact, two lines of breadcrumbs should lead us to the same conclusion.

First, there's a string of disease outbreaks—hard facts, if you will. Let's run through the widely publicized events; you will remember many of them yourself. From 2002 to 2004, severe acute respiratory syndrome, better known as SARS, affected some twenty-nine countries. Between 2009 and 2010, H1N1, or swine flu, affected many regions of the world, and the World Health Organization finally declared it a pandemic after hundreds of thousands of people succumbed to this flu. The Middle East respiratory syndrome, or MERS—caused by the MERS coronavirus—then occurred in 2012 (with additional smaller outbreaks in 2015 and 2018). And these examples are only some of the most prominent flu-related outbreaks (you might also remember the outbreaks of Ebola between 2013 and 2016 and of Zika between 2015 and 2016—both of them are viruses but not transmitted via air). Such outbreaks are not exceptions; they are part of life. So, in 2019 a new type of coronavirus emerged.

And then there were expert warnings—many. Most prominently, Bill Gates, the cofounder of Microsoft and the Bill & Melinda Gates Foundation, repeatedly warned of such a crisis. (A reminder that business celebrities indeed are valuable data sources, even for information outside their presumed area of expertise.) In 2010, he warned, "The H1N1 flu strain got a lot of attention in 2009. . . . The real story is that we are lucky it wasn't worse because we were almost completely unprepared for it."[4] In 2015, he underscored his concerns: "If anything kills over 10 million people in the next few decades, it's most likely to be a highly infectious virus. . . . We're not ready for the next epidemic."[5] The following year, speaking to Dame Sally Davies, the Chief Medical Officer for England, he put his finger on things to come, "We are a bit vulnerable right now if something that spread very quickly like a, say a flu that was quite fatal; that would be a tragedy."[6]

Government officials also highlighted the threats to the health of thousands if not millions of people and the global economy. In January 2017, longtime director of the National Institute of Allergy and Infectious Diseases, Anthony Fauci, who moved to the spotlight during the crisis, forewarned students at Georgetown University Medical Center. He cautioned, "There is no question that there will be a challenge to the coming administration in the arena of infectious diseases."[7] And at the Biodefense Summit in April 2019, the senior director on the US National Security Council for weapons of mass destruction and biodefense, Timothy Morrison, made it clear what his major concerns are: "When I was thinking about my remarks today, I pulled a book off my shelf, 'The Great Influenza'. . . . there were a couple of lines in here that ring true when I think about what keeps me up at night and what am I really worried about."[8] Finally, in September 2019—three months before the coronavirus first appeared in Wuhan, China—the US Council of Economic Advisers released its study aptly titled *Mitigating the Impact of Pandemic Influenza through Vaccine Innovation*.[9] There truly were enough warnings; there was no weakness in the signal.

In Timeline 11, you can easily see that actually two strands of developments establish not a weak signal but rather a very strong one. In fact, on the timeline the Covid-19 pandemic itself feels rather like another data point in a long line of developments than a truly unique marker of change.

No, the emergence of the coronavirus and the Covid-19 pandemic and its effects are not black swans, nor are they wild cards. In April 2020, in *Life after the Time of Coronavirus*, I indicated that such events will occur when billions of

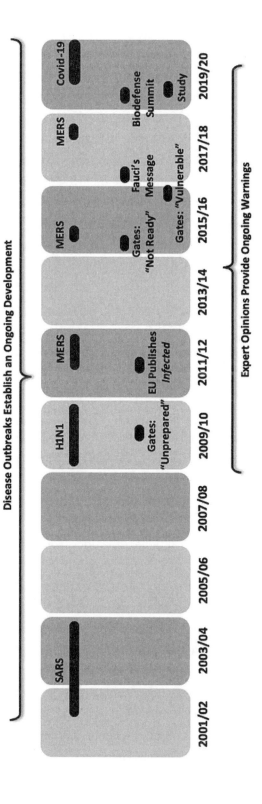

Timeline 11: The Coronavirus Pandemic's Strong Signals

people around the world are in close contact with increasing frequency.[10] The emergence of the coronavirus is similar to an industrial disaster or an earthquake: such events happen. Such events can be small or large; however, they all relate to the way the world operates in general. Similarly, this pandemic will not be the last one, and the world had better learn lessons from it. Rethinking risk-management practices and considering responses to future pandemics should be on the agenda of every policy maker and corporate leader going forward. Such rethinking— not the emergence of the pandemic—might be the actual signal of change that emerged during the crisis.

Unfortunately, the pandemic also put a spotlight on how decision-makers can spend a lot of resources and time on becoming aware of potential developments but then fail to prepare for such events. In 2011, the European Commission commissioned *Infected*—a discussion regarding how a pandemic could affect the world, which was published in 2012 (it's still online; go check it out).[11] The resulting comic book's narrative foresaw many of the situations and issues that would occur during the Covid-19 pandemic not even a decade later.

The comic book describes a traveler from the future who gets in contact with key people in the present to avoid the worst of an unfolding pandemic. He is trying to prevent the future from unfolding in the way he has already experienced it. The book features many similarities to the Covid-19 pandemic. The fictional disease is a zoonosis, an infectious disease that moves from animals to humans. In reality, the events were very similar; Covid-19 emerged from Huanan Seafood Wholesale Market, a wet-animal market in Wuhan, China. Then the story cautions how international airports will facilitate spreading such a disease to other countries. The comic book cautions, "Imagine if you were infected in this market by a new contagious agent. You probably wouldn't even realize it until the end of the incubation period." The storyline then continues to discuss measures to prevent the spreading of the disease, including people covering their faces, self-isolating, and avoiding use of public transportation. As a result, "Depression became the most common disease," according to *Infected*.[12] In fact, by April 2020, "more than a quarter of American adults [were] experiencing COVID-related symptoms of depression," according to a study by researchers from Boston University School of Public Health.[13]

The story has various elements of science fiction and crime literature. Many elements are purposefully fictional, such as time traveling. And many details of the story are very different from what would happen eight years later in the real world. But parallels between fiction and reality exist to a surprising degree. In fact, the story should have created awareness of what the main issues of a global pandemic later turned out to be.

But awareness and preparation are separate considerations. Clearly, decision-makers failed to heed some of the warnings that this work of foresighting surfaced. My team also outlined related aspects in 2017. In *Halt the Epidemics!* Ivona Bradley discussed the threat of pandemics, potential ways to predict spread, novel modeling of disease developments, and first efforts to enable rapid vaccine manufacturing.[14] The pandemic truly wasn't about awareness; preparedness was the missing piece.

But such events really are not part of regular scanning efforts. For one, a pandemic is simply not a signal of change; it's life. Let's say someone reported on the dangers of a major earthquake in the Bay Area in California—the region where I have lived for decades. "No kidding, Sherlock" would be the most likely response. Earthquake preparedness in the Bay Area is as much a part of business as usual as preparations for a pandemic should be on a global level. The question is how much effort and resources should be invested in advance for potential mitigation. Arguably, policy makers and many corporate leaders got the balance between preparedness and prudent spending wrong before the coronavirus took its journey around the globe.

Explore Novel Worlds

Despite wild cards that you should put aside for separate consideration, the use of the process for novel initiatives can focus your resources and efforts very quickly. Wild cards are a side note here and should see attention in a different managerial context.

The process of scanning can also find use for exploring entire market environments that are new to you. Let's say you understand that the Internet of Things could be an area you need to explore. You understand essentials; you might even have started an initiative. But it's a new area you have not worked with, a market you're unfamiliar with. You don't know what you don't know.

It is somewhat a reversal of the situation you will tend to encounter in that you are now seeking to find ways to initiate disruptions rather than identify and leverage them. The general task of looking for relevant data points and making sense of them remains the same. The need to develop a mental map of the area you want to engage with is a crucial task. And the impact/emergence matrix is such a mental map.

Scanning efforts that focus on a particular topic area—let's say the Internet of Things or autonomous robotics—can jump-start understanding of new business fields or market developments. As companies navigate increasingly connected marketplaces, the need for understanding of developments outside of their core competence will grow.

Adjustment of the process for such is straightforward, almost trivial. You simply need to restrict data points to events and developments related to these topics. Events might outline privacy considerations and security aspects of the Internet of Things or advances in materials of and consumer reactions to sensor-enabled devices. Naturally, ensuing discussions—and emerging patterns and narratives—that are based on these events will also relate to the topic of interest. Topic-related input will result in topic-relevant output. In particular, the impact/emergence matrix will then provide a handy overview of issues you likely will not have considered initially.

Interestingly, such a customized approach can also be used to explore the dynamics and effects of overlapping topic areas and market arenas just as, for example, by investigating developments in the Internet of Things and robotics at the same time. Particular topics might be fairly well understood by themselves; their interaction and dynamics might be hazy nevertheless. If you are active in both of the spheres, meaningful opportunities will emerge at the intersection. If you have strengths in only one of the spheres—let's say robotics but not necessarily Internet of Things activities—the process can serve to uncover competitive dynamics and identify partnership potential. In the case of investigating dynamics between the Internet of Things and robotics, you can easily see how increasing connectivity will affect robotics, in particular autonomous robotics, and how robotics can play a crucial role in Internet of Things applications.

Become Aware, Start to Prepare

Disruptions and change have become the norm: growing competition (increasingly from outside the industry), accelerating pace of technological developments, ever-faster shifts in consumer preferences, and proliferating business models all require strategic sensemaking. A restricted understanding of marketplace dynamics represents nothing short of a myopic view. A myopic view misses crucial junctures where markets move away from known pathways. A myopic view means setting the stage for failure. With a limited view, a company must ask not if but when it will find itself surprised by new business parameters that challenge its very existence. Such changes increase the importance of two managerial tasks. The first task involves finding the issues that open business opportunities or pose imminent threats. The second task is to understand their implications.

Becoming aware of the world is one thing. Assigning meaning to changes and disruptions is another thing. You can't do it all at once. The good news is that not everything has to be done simultaneously. And you can somewhat pace yourself by working off topics as you see them becoming relevant for your organization.

Identifying Meaning in a Nutshell

By identifying meaning, you put your organization into the context of outside developments. This step is the moment when you use your newfound knowledge of the future as a backdrop to your decisions.

When positioning topics of relevance, you are developing a map that is quite literally your view of the future. Because your company is the focus point of the map, you thereby took the first step from awareness to preparedness.

The map that emerges is the impact/emergence matrix. This overview will help you consider decisions you will have to make, and it will provide an order in which topics need addressing.

Impact highlights what is important for you. Changes are the norm now. Address the changes that matter.

Emergence illustrates uncertainties and timing attached to such changes. How speculative is such change? What is the timeline such change needs to unfold? It represents the roadmap of issues you will encounter as you move into the future.

The impact/emergence matrix is the sketch of the future that you will be able to share. Above all things, it facilitates communication of future issues. In particular, it will guide you in making the decisions you will have to make.

CHAPTER EIGHT

Take Action

GETTING A JUMP ON THE FUTURE

———————————————————■———————————————————

A good hockey player plays where the puck is. A great
hockey player plays where the puck is going to be.

—WAYNE GRETZKY

Kodak developed digital photography but got obliterated by competitors who
embraced the developing market. Nokia, the formerly uncontested number one
in phone manufacturing, got so tangled up in design and software issues that it
lost out to nimble and aggressive competitors. Apple, the largest company by
market capitalization, gained global dominance even though it did not create
the graphical user interface, did not develop music-file sharing, and entered the
phone market a full two decades after Nokia blazed its way to initial dominance.

By 2018 Kodak barely clung to life; Nokia had moved into the equipment
market, abandoning the consumer market almost completely; and Apple had
reached a market capitalization of $1 trillion (make that $2 trillion two years
later). What happened?

Kodak and Nokia failed to examine crucial market details and use them to
anticipate future possibilities. By studying the market at a seminal level, Apple
could look at the ways in which hidden market details would radically alter
the future.

However, timing is everything. Jump into the market too soon, and you can find yourself too far ahead of customers' needs and desires; jump too late, and, well, customers may dismiss you as a copy cat. The early bird gets the worm, but the second mouse gets the cheese. Timing is crucial; scanning's impact/emergence matrix helps you decide when to move aggressively and when to exercise patience.

Where the Rubber Meets the Road

You have arrived at the final step of the process. It is also the step that matters most. Filtering data points for manageability, identifying patterns and developing narratives, and prioritizing identified issues by considering implications all make you aware of how the world could change and what it means for you. However, you still have not decided what you will do now—now that you likely understand more about emerging changes and disruptions than many of your competitors. You're ahead of the curve, so make use of it.

Now you've reached a point that proves surprisingly dangerous for many organizations. If you don't turn awareness into preparedness, the efforts you've gone through could lull your organization into the notion that due diligence to anticipate the future has been conducted. Many companies stop here. Future explored, check. Tomorrow anticipated, check. But the first steps do not move organizations ahead. They show you a picture; they let you look at a mosaic of the future.

Although the importance of acting on new information is apparent, scanning efforts more often than not take place outside of companies' established silos; scanning in many organizations features unclear reporting pathways and can be transitory in nature. In many organizations, the real battle of the scanning effort is to drive relevant information and insights of emerging developments to appropriate functions and to the right decision-makers.

Turning insights into actions faces a timing consideration: you're ahead of the curve only for a limited time, likely a short time. And for many decision-makers, this point in time marks "an apparent paradox." Ansoff and McDonnell find, "If the firm waits until information is adequate for a decisive response it will be increasingly surprised by crises; if it accepts vague information, the content will not be specific enough for a thorough analysis and a well considered response to the issue." However, they provide a way to address such a strategic dilemma: "Instead of waiting for sufficient information to accumulate, a firm should determine what progressive steps in planning and action are feasible as strategic information becomes available in the course of evolution of a threat/opportunity."[1]

Such "progressive steps in planning and action" are the final step of FIPI: *initiate follow-up steps* (Figure 10).

Identify Meaning to Take Action

Developing an impact/emergence matrix very quickly makes sense of a wide range of developments. It's an overview of what you can expect to approach you in the future—or what you will run into.

Just as a quick reminder: we're looking at impact and emergence. *Impact* is attached in one shape or form to an objective: what are you trying to achieve? What is important to you? *Emergence* represents the timeline on which issues will unfold or how comfortable you feel with talking about what will happen. It's about timing and uncertainty. When will it become an issue for you? Is it already fairly understood what will happen? Or are we talking science fiction with implications that nobody can grasp today?

How do we identify emergence? As you develop your set of patterns and narratives, your team members can post them on the matrix. There will be discussion about what should be to the left, what should be way on the right. Those discussions will further your understanding, capture them. But don't attempt to achieve mathematical precision. We're talking about tomorrow. Seriously . . . we still cannot predict the future. But we can put issues in context. The relative position to each other counts. And you will find that agreement here is fairly easy to achieve. If not, well, then, you likely have a fairly uncertain topic at hand—almost by definition.

How do we capture impact? Here a simple voting exercise will do. Let your team decide what is important. Crowdsource your answer. Give each individual

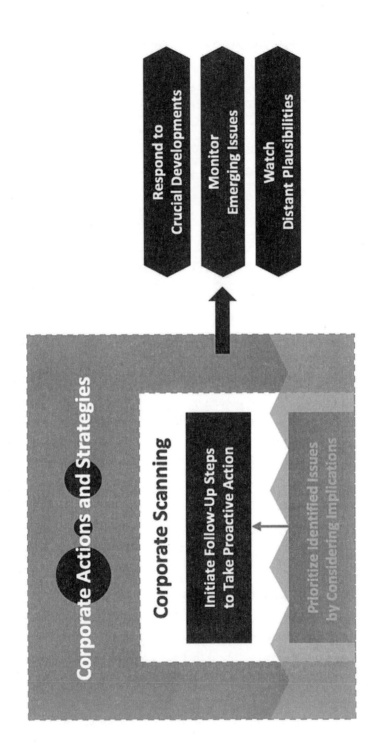

Figure 10: Step 4 Moves from Awareness to Preparedness

three to five votes and see what happens. Now make sure that everybody has only one vote for one topic. People tend to elevate the topics they see as their pet peeves. Also, don't let your team vote on topics you already work on. The assumption here is that you already find the issue important. Why would you have dedicated effort toward the issue otherwise? Such votes are lost votes. You're replicating efforts.

Schoemaker outlines a similar spectrum of decision-making situations that perhaps best clarifies the difference between data-driven strengths and narrative-focused approaches. His spectrum ranges from certainty, risk, uncertainty, and ambiguity to chaos/ignorance. He acknowledges that the earlier situations of the spectrum are "the most manageable and amenable to analytic approaches" but cautions that moving to other situations increasingly displays ambiguity that current tools and techniques are less suited for. Instead, "Here we must rely more on creativity and learning than on analytical deduction. The challenge in the middle is less one of computational complexity than of cognition, to make sure that we frame the issues correctly and ask the right questions."

In contrast, the topics to the right require a focus that "has to be more on generating multiple views, surfacing deep assumptions, and exploring the unknown terrain." Schoemaker also provides a rationale for why such a distinction is of increasing relevance for decision-makers: "Our modern challenge is that, overall, the world is moving to the right, with higher levels of ambiguity and even chaos in many sectors and industries."[2] Here we have a "test of a first-rate intelligence": to remember Fitzgerald's words, "The test of a first-rate intelligence is the ability to hold two opposed ideas in the mind at the same time, and still retain the ability to function." We don't know what will happen, but we should make sure that we keep alternatives in mind.

In the end, you will arrive at something like Figure 11—topics spread and sorted across the matrix. The question is: what to do with it? But you're already there; you might just not know it.

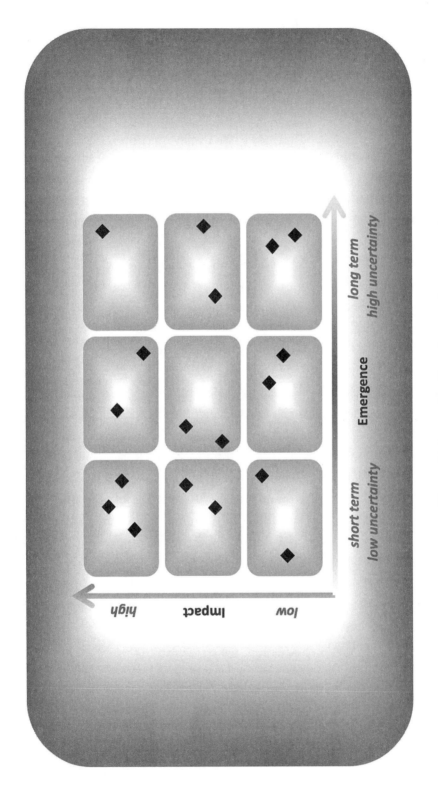

Figure 11: Assessing Developments

Approach Tomorrow—
Engage with the Future

The matrix is a shortcut to understanding the future. It offers an agenda that you can work off. Changes and disruptions in the commercial environment are conveniently presented. The framework offers an easily readable preliminary assessment of topics you should care about. It also offers you an intuitive way to communicate information and share insight. It is a versatile tool beyond scanning efforts to quickly sort through an otherwise overwhelming number of topics of potential interest or concern. Most important, the impact/emergence matrix is your guide on how to engage with the future (Figure 12). It shows you the pathway to prepare for the future. It literally provides the answers of how to deal with disruptions.

As you and your team search for memos from the future and develop patterns that point to disruption, you should also position the topics on the impact/emergence matrix. You should go through such an exercise at least quarterly. After all, the future doesn't stay still. But now you can see how the topics posted in the chart relate to three distinct suggestions illustrating how to respond to such changes.

The position of topics that received a threshold number of impact votes roughly will determine the response profile—the actions necessary to move you in a favorable position to face tomorrow. Clearly, only topics that bear significant impact on the company should find consideration. The chart is as much about calling attention to the most important changes as it is about discarding less relevant ones. To deal with uncertainty effectively, you need to make the future manageable. The short-, mid-, and long-term (or low, medium, and high uncertainty) scale then offers a scorecard about what to do with identified disruptions.

Respond to Crucial Developments. Naturally, some emerging changes will instantly strike a chord with your team. Hopefully, opportunities are obvious, but perhaps you notice an emerging threat that could endanger the very core of an organization's operations. Here topics immediately should be assigned to further research. You need to go deeper; you need to gain greater understanding about the nature of the change. Perhaps you have found a crucial juncture of the future for your business, your industry, perhaps even the global economy.

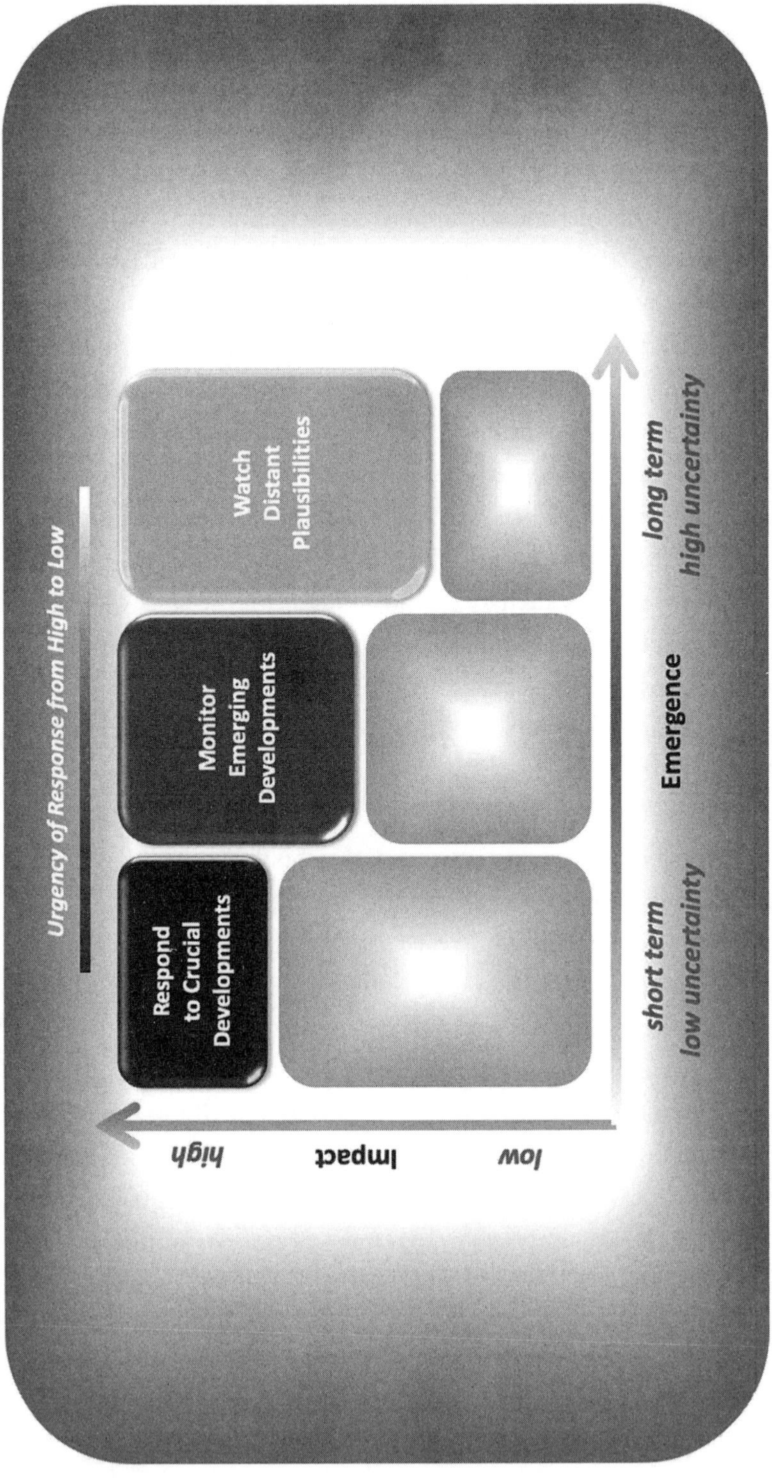

Figure 12: Future-Oriented Actions

Reaction time matters then. *Responding to crucial developments* requires immediate attention; here the awareness gap can be your ticket to the future. You indeed might know something now that most are unaware of. Don't get carried away; analysis might show that the change will take a long time or that its relevance for your organization is not as impactful as you initially thought. But immediacy in dealing with the change is crucial. It is here that competitive advantages are made—or wasted.

Monitor Emerging Developments. Other potential disruptions will resonate with your team. These changes might pique interest, but without leaving a feeling of immediate need for action. Such topics could become problem areas or opportunities. But time needs to play out; other conditions need to emerge. Tiles still have to fall into place. Brushstrokes are still necessary to complete the picture. They provide an early alert that an adjustment in strategy might be necessary some time down the road, but currently they are simply too immature to justify any substantial resource allocation. Put some options in place; add them to potential changes you are already monitoring to ensure that you stay in touch with the changing fabric of the marketplace. But stay alert. You uncovered potentially game-changing disruptions that could emerge if the petri dish of the marketplace provides the right conditions. The changes could come to full fruition with only a number of additional building blocks falling in place—a regulatory decision, a key player's participation, or an enabling technology's commercialization. Keep tabs on related developments. Don't squander your awareness lead.

Watch Distant Plausibilities. You are looking at the future. Uncertainty is your constant companion. You will become aware of many potential disruptions that could occur; after all, creating awareness is the point of going through the first part of the steps that FIPI outlines. That does not mean that everything you become aware of is ripe for action. Many developments feature deep uncertainty. You might not know if other developments will accelerate the change or if friction areas will hold it back. Too many stumbling blocks and curveballs make it difficult to envision what the effects of this change will be, if there will ever be disruption. Anticipation here is limited. Too many tiles are still missing to tell you what the final mosaic could be. But you found potentially crucial issues. They just don't justify getting bogged down or distracted; more important and immediate changes require attention. Instead, such topics should find their way onto a watch list—a list of issues, concerns, and considerations that could affect you dramatically—just perhaps not tomorrow, but maybe the day after tomorrow. Topics on such a list should find frequent revisiting to determine if related changes have occurred. Again, don't waste a lead.

The impact/emergence matrix provides you with the focus to react to changes at the right time with the right involvement. By frequently using this matrix, you will fill in the canvas of what tomorrow's game-changing developments might be; you will develop a feeling for riptides and undertows that lie underneath every market's often calm surface.

Over time you will gain a visceral perspective of the ebb and flow of issues in the marketplace. Some of them—to stay with the metaphor—will become flotsam that changes position somewhat and sometimes rises just to then vanish into the background again. Privacy issues have been at this stage for decades without genuine impact on the marketplace. After all, big data would have a much harder time collecting all the information. Facebook would not be worth more than $800 billion, as it was in October 2020, if privacy were a major consideration. But be careful, flotsam can arise to become an issue: the European General Data Protection Regulation might have been the crucial watershed moment.

Other issues will show persistency. They frequently will come up, call your attention, and over time preoccupy you. These issues might be the disruptions that change the business world for good. Social networking applications and services clearly have come a long way from their onset perhaps in the late 1990s or early 2000s and expanded their commercial reach from pastime activity to one of the most prevalent communication activities to becoming powerful enterprise and government tools. The Internet of Things has moved from a theoretical and academic concept to an all-encompassing buzzword that mobilizes billions of dollars of investment.

Other strategists have their own recommendations, but they generally highlight the same pathway forward. Ansoff and McDonnell consider three categories of interest. "(a) Highly urgent issues of far-reaching effect which require immediate attention," "(b) Moderately urgent issues of far-reaching effect which can be resolved during the next planning cycle," and "(c) Non-urgent issues of far-reaching effect which require continuous monitoring."[3] Day and Schoemaker describe corresponding responses in much more intuitive ways: believe and lead, probe and learn, and watch and wait.[4]

But here is a warning. Responses to early understanding of change can be hesitant. Organizational behavior scholars Karl Weick and Kathleen Sutcliffe urge, "The overwhelming tendency is to respond to weak signals with a weak response. Mindfulness preserves the capability to see the significance of weak signals and to respond vigorously."[5] A weak signal provides a competitive advantage only if you act strongly on it. The awareness gap offers a lead only if you grab your newfound insights and run with it.

Now all the steps of FIPI have come together. You know now how to develop awareness; you understand how to prepare yourself. The entire process is shown in Figure 13. Simple but powerful steps let you leverage small data to find the big disruptions that will shape tomorrow. Now let's put together all the pieces.

The entire process—FIPI—serves only one purpose: to position you for the future. Get in gear; move to the right spot. Play where the puck is going, not where it is. Be a great player.

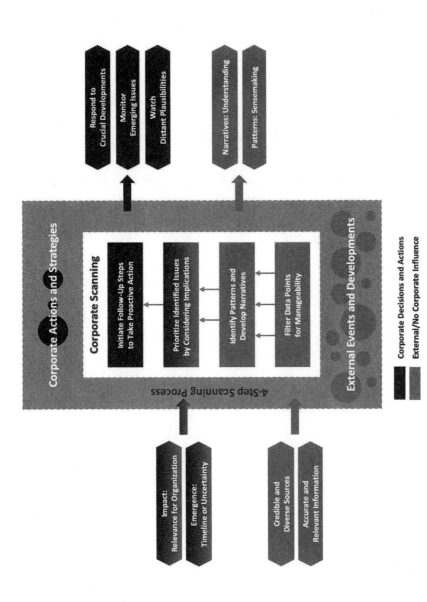

Figure 13: Complete Overview of Scanning

Taking Action in a Nutshell

Initiating follow-up steps is the crucial part of FIPI. Without this last effort, everything you did before will stay meaningless. It's time to move from awareness to preparedness.

....

The impact/emergence matrix offers you a convenient way to segue from considerations to actions. The matrix provides you with a shortcut to decision making and a sketch of the future that is easy to communicate.

....

Responding to crucial developments addresses the most urgent concerns. In fact, such developments might represent existing organizational blind spots that already threaten oper-ations. Move on them—now!

....

Monitoring emerging developments alerts you to issues you need to keep tabs on. They alert your senses. They let you move into a starting position to respond rapidly as the need arises.

....

Watching distant plausibilities keeps your antennas up. Develop a watchlist of issues that are still too amorphous or uncertain to make meaningful sense of them. But be aware that something is brewing. Perhaps you shouldn't respond right away, but now that you know, it won't be a surprise anymore.

....

Finally, don't make FIPI a one-time exercise. It's an ongoing process. Keep your forward-looking senses sharp and your organizational muscles flexed. The future won't stop; you shouldn't either.

....

CHAPTER NINE

Abandon Today

FIGHTING FORCES THAT KEEP YOU IN THE NOW

———————————————■———————————————

Culture isn't just one aspect of the game; it is the game.

—Lou Gerstner Jr.

Companies regularly miss crucial developments. Big companies miss big developments. And category leaders become has-beens overnight. Worse, opportunity knocks on your door until it passes to the neighbor's house because you did not promptly respond to its invitation. In 2000, opportunity invited Blockbuster to purchase Netflix for $50 million.[1] The door stayed shut. And it did not open for Yahoo! in 2002, when it balked at adding $3 billion to its offer of $2 billion to buy Google.[2] (To add insult to injury, Yahoo! also had turned down Google's entrepreneurs when they tried to sell the company for a mere $1 million in 1998. Opportunity rarely knocks twice.) Blockbuster filed for bankruptcy protection in 2010; Yahoo! ended up as a mere unit of Verizon Media's Oath division. Netflix? Worth over $200 billion in 2020. Alphabet? Over $1 trillion, thank you very much.

But 20/20 hindsight makes these sorts of mistakes seem all too predictable. Why didn't Blockbuster and Yahoo! answer opportunity's knock? Four main reasons account for companies' shortsightedness: short-term focus, quantitative obsession, and the selected organizational home; most important, however, corporate culture is the main hindrance to anticipating the future. After all, culture is at the core of shortsightedness no matter the type. It is unfortunately at the fabric of each company. Changing culture is unraveling fabric and restitching it. The wrong culture creates myopia; the right fabric establishes competitive advantage.

Be Aware of Hurdles to Anticipation

The merits of scanning are straightforward; the process of implementing scanning is not. Inertia can be a very obvious challenge. The assumption that forward-looking efforts already exist—somewhere—within an organization is a more common hurdle. Structural considerations play as much a role as attitudes and the corporate culture. Where to begin? Being aware of common challenges is a first step.

Figure 14 illustrates the four major hurdles of a mindset of anticipation, a habit of living in the future. These hurdles are powerful barriers to scanning. These hurdles isolate decision-makers and their organizations from markets and emerging developments. They glue organizations to today. They are the stepping stones to future failure.

I'm not saying that information from the future is completely absent. Organizations are not blindfolded, but they might not see far and wide. The challenge is to bring the pieces together and to put them into a comprehensive mosaic—or to let all the brushstrokes paint the full picture. Bazerman and Watkins find that "different divisions of an organization may possess separate pieces of the puzzle, while the organization as a whole lacks the personnel and systems necessary to integrate and distill them into actionable insights."[3] True, but organizations need to overcome a number of hurdles—hurdles that have become higher over the past decade.

The Focus on Today Is the Blindness for Tomorrow

Time is money. Short-term returns are in high demand. There's no question that patience has become an increasingly elusive virtue. "Short-termism is often cited

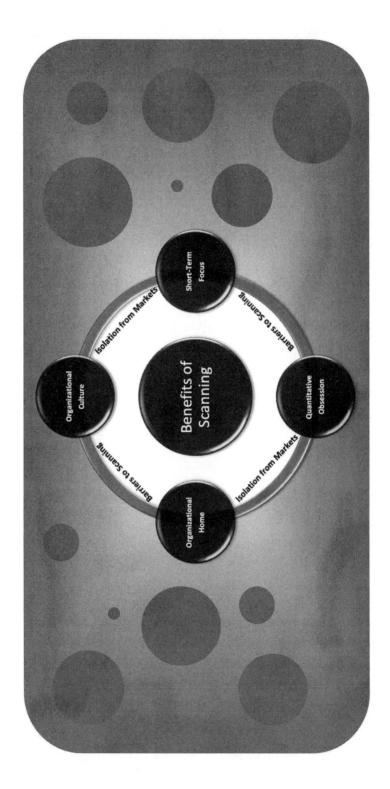

Figure 14: Implementation Hurdles

as a barrier to traction with horizon scanning outcomes," according to the UK government's review of its Horizon Scanning Programme.[4]

Short-term thinking comes from different rhymes and reasons. It's worthwhile to untangle the different flavors. Some are of internal making, some are external obligations, some are organizationally driven, and some are market determined. External pressure, the requirements of the financial community, has built up expectations that can prevent a long-term, strategically disciplined approach. Investors have come to take quarterly reports as measures of success. The financial market operates under different constraints and timelines than companies in their respective industries should operate under. Friction between short-term returns and long-term operational success is a very real concern decision-makers have to balance.

In a 2013 management survey by researchers at McKinsey & Company and the Canada Pension Plan Investment Board of more than one thousand high-level decision-makers, almost two-thirds of them admitted that pressure to generate short-term financial results had increased since the 2008 financial crisis. Some 79% mentioned that they feel pressure to show strong financial outcomes within a period of only two years or even less. Crucially, "86% declared that using a longer time horizon to make business decisions would positively affect corporate performance."[5]

Decision-makers can extricate themselves from such pressures. Very prominently, Warren Buffet has led Berkshire Hathaway for decades with a very long-term perspective. Other companies have made the point at times that a quarterly orientation and frequent earnings guidance cannot be in the best interest of the company and therefore investors. The formation of Alphabet, a holding company, enabled Google's founders Larry Page and Sergey Brin to separate the profitable core of their operations from speculative endeavors such as X and GV, the company's R&D and venture-capital subsidiaries, respectively. The separation split short- to mid-term and long-term considerations, thus reducing conflicts of interest.

Internal issues exist too. Corporate decision-makers are also individuals. Many high-impact managers and executives do not expect to stay long with a company; creating short-term success opens career opportunities. Short-term contributions can be very visible and assignable to individuals. Long-term efforts develop slowly, and many hands will be associated with the longevity of an organization. Individual ambition and corporate strategic needs tend to be misaligned; individuals' focus on short-term metrics endangers the long-term strategic awareness and success. Bazerman and Watkins similarly point to misalignment of incentives. "Notably, predictable surprises often play out over time frames substantially longer than the typical tenure of organizational leaders."[6] Managers then will come to the conclusion, "Better to focus on my short-run goals and reap the rewards."

Such personal interests can sabotage long-term thinking at the core. I'm not saying that individuals set out to do so. I am saying that their personal interests result in organizational myopia. Bazerman highlights such a conflict between personal and corporate interests; he even gives it a name. *Motivated blindness* occurs when people are incentivized to ignore potential problems and do not attempt to communicate related issues for personal reasons.[7] Motivated blindness requires a fundamental look at the organization and incentives. Such conflicts occur on every level along the entire hierarchy of your organization, so be aware. Scanning can be a way to institutionalize a long-term view, to make short-term incentives more apparent.

Misguided incentive schemes can steer individuals purposefully or accidentally into short-term actionism that overrides long-term sensibilities. The subprime mortgage crisis in the US between 2007 and 2009 provides an illustrative case in point. Individual lenders were incentivized to sell subprime mortgages and personally benefitted financially. Companies had incentives to develop ever more arcane packaging of loans to boost selected metrics. In the end an entire industry—the global economy really—was moving toward a cliff. The pain of short-term thinking can ripple through industries and economies. Here a wide range of players, even policy makers, fell victim to the allure of short-term gains. Day and Schoemaker call attention to the concern, cautioning, "Strong focus of attention may benefit short-term performance but may work against the organizations' long-term survival, particularly when the environment changes."[8]

Likewise, short-term focus breeds short-term needs. Many organizations never developed awareness of the future. They simply missed crucial signals of change in the past. Now they are too preoccupied with dousing current fires of past blindness. Steve Tighe, a strategist who used his own past professional experiences and disappointment as a platform to advise other executives, now warns of an attitude that "translates to: 'We're too busy dealing with today's realities to worry about tomorrow's possibilities.'"[9] And such preoccupation comes at the cost of missing the next signals of change. A vicious cycle. This is where "I didn't see that coming!" originates.

Obsessing over Numbers Hides Warning Flags

Scanning faces strong headwinds in the time of big data. Qualitative understanding of the world seems quaint when artificial intelligence delivers analytics for every taste and flavor—conveniently without the need to think through dynamics. The problem here is that results reflect past dynamics. Big data and analytics require a lot of information—information that already is abundantly available, information from the past, often lagging indicators such as sales or cost developments. But "a lagging indicator is an outcome or consequence of some activity that came before," as McGrath reminds us. Instead, she highlights the power of "leading indicators [that] are often qualitative rather than quantitative. They are often told as narratives and stories."[10] New developments don't start with numbers; they start with events. The seed of something truly new is a single activity, not a percentage, a share, or a growth rate. Those metrics indicate that something is already in full swing. Such metrics let you discover the past. Worse, they tend to hide the future.

"But aren't there trends?" you ask. Indeed, and they deserve attention. But they are continuations, not innovations. They represent existing momentum, not new pushes. They shouldn't be ignored, but overemphasizing trends only makes you part of the herd—one of many whose strategy is a dime a dozen. Steve Tighe recounts a personal experience as an executive who analyzed presumably indicative trends to then find himself following the crowds: "Trends don't provide competitive advantage. [Trends'] visibility provides a guide for planning and innovation. But this visibility is also a weakness for organizations, because . . . competitors were doing the same." The outcome is far from innovative behavior: "The resulting convergent behavior that trends promote is akin to moths being attracted to a flame."[11]

I'm not arguing against the use of quantitative metrics. Decision-makers require guidance about how much input resources will cost, how many widgets will be sold, or how much consumers will be willing to spend on this year's holiday shopping. No forecast will ever be right, often not even close. Whereas such forecasts look at understood dynamics, scanning is the attempt to develop an understanding of future marketplace realities. Scanning is about identifying and learning to understand emerging new dynamics. Weak signals instead of big data. Forecasting and predicting live in very different corners of organizational sensemaking than does scanning. Scanning and these approaches do not have to compete but can be powerful partners. But "bad forecasting . . . steers us subtly toward bad decisions and all that flows from them—including monetary losses, missed opportunities, unnecessary suffering, even war and death," as Tetlock and Gardner highlight.[12] I'm arguing that qualitative information is underrated—particularly when algorithms and machine learning seem to have all the answers. Scanning is rarely in competition with other methods and processes but rather complements such efforts.

I frequently encounter the submission to quantitative approach or computer-driven approaches—or is it the resistance to trust human ingenuity? I frequently face questions from workshop participants wondering why an algorithm wouldn't be able to do the same. Is it the belief that machines are smarter, or is it the hope for convenience instead of making one's brain cells work? Big data promises are fanned by an industry that lives on the perception of algorithms as a panacea. Misguided hopes are to blame too. In 2008, author Chris Anderson stated in the now seminal article "The End of Theory: The Data Deluge Makes the Scientific Method Obsolete" that "with enough data, the numbers speak for themselves." After all, "faced with massive data, this approach to science—hypothesize, model, test—is becoming obsolete." In particular, he argues, "No semantic or causal analysis is required."[13]

Anderson tried to make a point. I get it. But the result of such thinking over time has become an obsession with the power of the approach. A dangerous obsession, I argue. Semantics matter and causation provides investigatable dynamics. Semantics not only define the problem (as well as the solution) but also make problems tangible. Numbers can lack human relevance. Numbers rarely tell a story. Numbers don't provide underlying motivations. And causation pinpoints the problem—as well as pathways to solutions. Numbers are tools to address semantics and causation, not to do away with them.

Campbell Harvey, professor of finance at Duke University's Fuqua School of Business, points to a more general issue: "Unfortunately, our standard testing methods are often ill-equipped to answer the questions that we pose."[14] Mathematical obsession has ethical dimensions too: human decision-makers will be robbed of understanding and therefore control of the process, application, and results. Randomly assigned to a handful of personalities and phrased in different ways, the quote "Not everything that matters can be measured, and not everything that can be measured matters" still provides an insight into practical relevance.

Quantitative data can become an addiction among decision-makers. Data-driven operations can become corrupted by the lack of desirable data but the availability of a wealth of measurements of less relevant metrics. For businesses, strategic blindness emerges; in war, lives are at stake. Army advisor James Willbanks observed how data collection became an obsession and how nonrelevant metrics were used to justify strategies during his time in the Vietnam War. He recounts, "The problem with the war as it often is are the metrics; it is a situation where if you can't count what's important, you make what you can count important."[15] Essentially, the tail is starting to wag the dog.

Operations during the Vietnam War are illustrative. Lessons could be learned but somehow were lost. During the time of the United States' involvement in the war, Rufus Phillips was with the United States Agency for International Development (USAID). He relates a story about Robert McNamara, who was Secretary of Defense from 1961 to 1968 during the crucial phase of getting directly militarily involved in the war. McNamara is considered a pioneer in the field of systems analysis.

Phillips remembers, "Secretary McNamara decided that he would draw some kind of a chart to determine whether we were winning or not. And he was putting things in like numbers of weapons recovered, numbers of Viet Cong killed. Very statistical. And he asked [US Air Force officer] Edward Lansdale [at the time assistant to the Secretary of Defense], who was then in the Pentagon as head of special operations, to come down and look at this. And so Lansdale did, and he said 'There's something missing.' And McNamara said, 'What?' And Lansdale said, 'The feelings of the Vietnamese people.' You couldn't reduce this to a statistic."[16] The crucial piece of information was lost.

Hans Rosling provides a general assessment, "It's not the numbers that are interesting. It's what they tell us about the lives behind the numbers."[17]

Not only did the Vietnamese perspective get lost, but the US military lost sight of its own population's sentiments. (Losing sight is at the core of concern when relying on numbers.) The narrator of the documentary *The Vietnam War*, Peter Coyote, comments on developments regarding the statistical rationale of the operations during the Vietnam War: "When [US] Senator Fritz Hollings [from South Carolina] visited Saigon shortly after the Ia Drang battles [in 1965], General [William] Westmoreland [commander of the US forces during the Vietnam War from 1964 until 1968] told him, 'We're killing these people at a rate of 10 to 1.' Hollings warned him, 'Westy, the American people don't care about the 10, they care about the one.'"[18] Math can be funny. It can be perfectly correct, defensible in every boardroom—that's why it's used. And, at the same time, it can completely miss the crucial insight, the aha moment—that's why its use should be reevaluated.

The situation is bound to get worse before it gets better. Artificial intelligence and deep learning, themselves technology enablers for a wide range of applications, depend on massive amounts of data. Nothing wrong here if such enablers are used for the right problem by the right people—likely that won't be the case though. Too strong is the allure of effortless machine-driven decision making. Society is fascinated. Some decision-makers are willing to abandon common sense. Economist and journalist Tim Harford suggests, "Big data is a vague term for a massive phenomenon that has rapidly become an obsession with entrepreneurs, scientists, governments and the media."[19]

The issue is an obvious one. Big data is about a lot of data, obviously. Good for operational tasks; potentially disastrous for strategic sensemaking. Commercial disruptions emerge from a handful of developments, even single events. By the time big data is available, the news has become history. By the time machines can crunch the data, decision-makers have already missed the boat. In fact, genuine breaks with tradition and convention tend to be driven by singular events. The first email sent, the first time graphene was created, and the first internet-connected robot provide vital clues for emerging pathways to new societal behavior, promising material-driven changes and new manufacturing paradigms. Disruption needs a push that tips over the first domino to catch the imagination of the marketplace.

Scanning relies on the human strength of filtering crucial bits of information out of the sea of random data points. Scanning provides insights by humans for humans. More important, such efforts establish narratives—narratives that can be easily conveyed and also very easily understood by decision-makers. Narratives elegantly deal with conflicts and contradictions in developments and dynamics; data analytics throws out confusing gibberish. Flaws and fallacies in narratives stick out like sore thumbs; decision-makers intuitively see the dangers.

Data analytics buries genuine newness, hides true disruptions. Amar Bhidé, professor at Tufts University's Fletcher School, argues, "Statistical models disregard the uniqueness of events, treating them like balls in a jar that vary only by diameter or color."[20] The "uniqueness of events" is at the very core of scanning activities; these events represent the needle in the haystack.

Missed needles can be disastrous, and sometimes everyone hurts. Statistician Nate Silver argues the dangers of the US housing crash of 2007 and 2008 actually had been identified and reported on by various media outlets well before its occurrence.[21] Then rating agencies applied number models. The institutional need to work with quantified information masked the urgency of the situation. Silver also cautions that more data will result in more noise and misleading or confusing bits of data. The signal—the needle—can be drowned out in a sea of data. Taleb, who investigated the impact of rare and unpredictable events, cautions, "Modernity provides too many variables, but too little data per variable. So the spurious relationships grow much, much faster than real information."[22] Meaningless correlation likely will become the norm; a true game changer will have a chance to hide among such correlations until it's too late.

Scanning is not about calculating correlations, but about unearthing signals of change of developments that affect markets on a much higher, all-enveloping level. Correlations, extrapolations, and predictions prove self-limiting, putting the focus on one possibility instead of exploring the full range of options. For corporations such an approach reduces the ability to distinguish their strategies from competitors', eliminates the potential to find truly groundbreaking strategic pathways, and perhaps increases the operational and executional risk. After all, all numbers being equal, a corporation then will find itself in a pack of competitors vying for the same resources, distribution channels, and consumers.

Martin provides common sense in a world of algorithmic focus: "The world is not responding to our attempts to control it with quantitative models. Our chaotic environment demands a new approach that pays attention to qualities in addition to quantities."[23]

Home Is Where the Heart Is

Like weak signals are the outliers in data sets, scanning is the outlier in corporations. Professionals don't know what to make of it; decision-makers don't know where to put it. But location is crucial. It needs the right support and the right channels—support to make identification possible, channels to enable communication. Not finding an appropriate organizational home can lull decision-makers into feeling they've done due diligence without any results to show for their efforts. "Location, location, location" doesn't just apply to real estate.

Some companies rely on their R&D, strategic-planning, and marketing departments to surface meaningful developments early. They assume that miraculously such information will find its way to the right team in the right department or function—a team that then knows what to do with it. Then again, the situation can get worse. Other corporations are so focused on their current business that they essentially leave this task to the attentiveness of individuals within the organization: somebody somehow will feel responsible to identify and communicate crucial information. Day and Schoemaker point to the main issue: "In many organizations, the periphery is everybody's responsibility—and nobody's."[24] If leadership does not perceive the importance in establishing the necessary structure to facilitate scanning efforts, it is difficult to see why individuals would take it upon themselves to struggle through the thicket of organizational hurdles.

Challenges any scanning process faces broadly fall into three categories. First is the issue at hand. Disruptive changes in the marketplace *are* difficult to identify. And that's not all. If the importance of changes is misjudged, nobody will react. If identification doesn't occur in a timely manner, everybody will react—only in a frantic fashion. Second, companies—particularly successful ones—tend to develop ignorance toward external developments. Ongoing operations take precedence; R&D and product development override contradictory signals. Third, as companies grow, structures within become too complicated to communicate in effective ways. Success and size make companies lazy and complacent.

Location of activities within an organization is an important consideration for every function. For scanning, it's a make-or-break decision. Don't bother if you're not willing to consider a good home for it. It will be difficult for a product manufacturer to disregard logistics. But it's not uncommon that an organization can virtually forget that it has established a scanning effort—if located in a barely noticeable corner of the organization. Where such efforts are organizationally positioned determines perceived importance of an activity. Its home will decide how much attention staff will pay to the efforts. Its home shapes communication of insights.

Involvement of decision-makers is crucial. The insight is not new. Not many companies follow the guidance though. The first serious discussion of horizon scanning already highlights the need. Aguilar stresses the need to get senior management involved: "Coordination through explicit assignment must start with the men [and women] at the top, the company's most influential men [and women]."[25] Ansoff and McDonnell also see the challenges and importance of getting decision-makers and managers from different functions on board. In addition, they stress another consideration: "Strategic and creative managers must be 'sold' not so much on the weak signal concept as on the ideas that systematic management of weak signals is desirable."[26]

Finding a good home for scanning operations sends a message to staff in two ways. First, staff members simply need to know that such an effort exists; visibility within an organization is important. Second, where scanning is located informs about the perceived importance. Staff will use it as a marker to deduce how seriously to treat the effort. The wrong home thwarts any good-meaning effort.

The ability to communicate results is the crucial aspect of any information-gathering effort. Surprisingly, in scanning, this aspect turns out to be a prerequisite that is prone to be neglected. Some companies treat scanning efforts as boxes that need to be checked. Other companies assume that scanning participants will take it upon themselves to spread the message. Work and time go into the effort; results stay hidden or with a selected few.

In general, I have encountered three different ways companies attempt to meaningfully integrate scanning efforts within their organization. First, companies that see themselves as reactive—rather than driving—market participants tend to place scanning efforts within the company's existing market research. It's another source of information. Second, organizations that see operational excellence as their crucial commercial contribution will attach scanning to operational tasks—for example, manufacturing or marketing—to enable ideas to emerge where they are needed most. Scanning therefore becomes a scouting service for technologies and approaches that focus on a specific task. Third, firms that have a keen eye on innovation will place the scanning function closer to strategic planning and leadership. Scanning becomes a lifeline for future success.

All three approaches feature their own benefits and establish their own challenges. Attaching scanning to existing market research and intelligence provides the effort with a clear place in the organizational structure that features understood communications channels. Results are easily conveyed; narratives find readers. Potentially, though, scanning—as a qualitative approach—can find itself in competition with data-driven approaches; worse, data scientists can view the qualitative approach as inferior to quantitative analysis. Scanning then will suffer because of the constant need to vie for attention with a slew of other market-research and analytical approaches.

In comparison, if scanning is embedded in operational day-to-day activities, in many cases the information emerges where it is needed and where it can be put to action most quickly. The need is understood, attention will be paid. But day-to-day operation is a very different animal than strategic considerations are, and scanning might suffer neglect, so it is easy to see how staff will treat the efforts as an afterthought. In a best-case scenario, scanning becomes part of a contained effort; in the worst case, results end up in a drawer or efforts fall asleep and enter a very deep slumber.

In contrast, placing scanning with strategic planning and leadership elevates its impact. Hurdles to implement such an effort are high; a considerable amount of cost-benefit thinking occurs before the process is implemented. Buy-in to the concept, understanding of strategic benefits, and signaling of organizational relevance are needed from senior management. A lot of education and conviction has to be in place. Once established, though, the process has the best chances to exert a strong influence on corporate strategy and market conduct. Insights will flow from the top down, encountering fewer hurdles than the other way around.

Home is where the heart is. Organizational considerations are one issue; staff members' focus another one. Day and Schoemaker caution: "Groups such as corporate development, competitor intelligence, market research, or technology forecasting can be given the task of scanning. The risk is that these mid-level groups may limit themselves to collecting and processing data from the domains they know best rather than scanning broadly and educating others about what they have learned."[27]

Bazerman and Watkins emphasize two crucial failures looking at the future—failures that the right implementation of a scanning system can address. *Scanning failures* are "failures to engage in adequate scanning of the external and/or internal environment, either due to lack of resources or to organizational inattention regarding important classes of threats." Most organizations are aware of that concern—although they might not act on it. As problematic are *integration failures*—"failures to put together disparate pieces of information possessed by various parts of the organization, or to analyze available information into actionable insights."[28] The right home with the right channels prevents both failures. Time spent in finding the right place is an investment in a conduit to the future.

Culture Is the Game

The previous hurdles are issues of a bigger problem. Hurdles can be overcome, but culture is the builder of such hurdles. Culture is at the center of all things standing in the way of your understanding of the future. Lou Gerstner Jr., CEO of IBM from 1993 to 2002, found, "Culture isn't just one aspect of the game; it is the game." He's right.

Scanning does not exist in a vacuum within organizations. Corporations have divisions, departments, and teams scattered across geographies and continents. Such fragmentation is a function of departments' and countries' specific needs. Such fragmentation poses very fundamental questions: Where should any scanning efforts be located? And what are practitioners' expectations of the process and benefits?

More often than not, the decision to implement scanning comes down to individual decision-makers' personal desire to look into the future and to embrace uncertainty. How to anchor the process within an organization then is a matter of personal preferences rather than corporate requirements. A corporate culture's openness and catering to forward-looking efforts are the crucial enabler. Weick and Sutcliffe maintain, "Culture affects how we detect, interpret, and learn from disrupted expectations."[29]

Culture shapes perceptions. Staff can easily mistake the newfound capability as a challenge and threat to their own work. But scanning should not be perceived as competition because it relies on, even promotes, the already-existing resources to make sense of the external environment. Nevertheless, managers have to be prepared that groups or researchers can get combative. It is not uncommon for me to see R&D managers questioning the process and results when I present patterns and narratives from the future. After all, shouldn't that be their job?

Another type of competition is even more confrontational: the competition for time and attention. Some planning and research groups constantly switch from one internal project to another—often addressing fires rather than following deliberately laid-out planning efforts and research schedules. Professionals will have a very hard time freeing any additional time to support a process that still looks for acknowledgment and support. Also, projects tend to be short- or mid-term affairs—in short, a promising way to showcase individuals' abilities and put them on the map. Scanning, in contrast, is geared toward long-term results in terms of knowledge gain; the glamour and glitter can be missing. Promising short-term achievements (organizationally as well as personally) vie with long-term benefits—again! Personally, such a view is not unfounded; many organizations indeed prioritize short-term high-profile projects. Organizationally, such a culture promotes myopia. Executive levels have to show commitment to scanning if they intend to make the process successful.

Then there are realities—realities that are the result of past negligence. Organizations that face a constant onslaught of emergencies—or perceived emergencies—are distracted. An ongoing string of short-term actions prevents organizations from looking at emerging issues. Emerging issues become new realities. These realities then require short-term actions. A vicious cycle is born.

Instead of putting out fires, management instead should identify embers early to prevent such fires. Arguing that ongoing operations have to take priority over any consideration of speculative futures is a crucial fallacy. Current realities, emergencies really, came from past futures that remained unidentified. The lack of long-term perspective that scanning could provide often is a reason for putting out seemingly continuous flare-ups. Simply put, the organization has become so focused—so myopic—that virtually every change in the environment comes as a surprise. Scanning could guide such organizations to a less frantic, more prescient way of managing the future's opportunities and threats.

A Lesson in Complacency

Many companies' fates can illustrate a lack of strategic awareness very well, such as those exemplified by AOL, Blockbuster, Kodak, and Nokia. Entire industries that were seen as unassailable, such as the US steel industry, the British automotive industry, and the German consumer electronics industry, similarly succumbed to new realities that were not on their radar. Today, many energy providers and most incumbent car manufacturers face the threat of a value-creation maelstrom of environmental necessities, advanced technologies, and consumers' changing tastes. Electric propulsion, driverless technologies, and a proliferation of infotainment and safety components have opened doors for electronics and software companies that carmakers simply didn't see coming at first.

Then there is psychology. A lack of quantitative analysis in qualitative assessments can make decision-makers very uncomfortable. Therefore, forecasts and predictions of market sizes, unit-sales developments, or regional market growth have become sole focal points of decision making. Quantitative information can only reflect known and understood economic behavior and commercial conduct. Untested business models, emerging consumer behavior, disruptive technologies, and novel application areas are nearly impossible to capture in numbers and trend lines. Early detection by its very nature means qualitative detection; numbers then can offer guidance or verify existence of new developments. Worse, psychology can make thinking of change challenging. Bazerman and Watkins spell out related hurdles, the "Natural human tendency to maintain the status quo." They continue, "Most people will discount the need for fundamental change, at least until a tragedy occurs. In addition, people tend to ignore errors of omission. That is, we tend to ignore the harm that occurs from inaction and pay greater attention to harm that comes from action."[30]

Nokia represents a case study illustrating how overwhelming success can beget its own strategic dead end. Without doubt, the company was the leader in the cell-phone industry. Also without doubt, it has seen its fortunes wane astonishingly quickly. Personal experience provides anecdotal evidence regarding how success can breed a culture of discounting change. For years, Nokia had used scanning approaches—internally and with help from my team—to provide its decision-makers with an overview of emerging developments. I had the chance to work with the company repeatedly on its WorldMap, a scanning effort to identify the crucial developments that would impact Nokia over the next half a decade. Between 2003 and 2006, the company took efforts to design such lists of developments and followed process and structure as described in previous chapters. In the early 2000s, Nokia also had a group in Silicon Valley: "Innovent, an entrepreneurial innovation unit in Nokia, connects to the pulse of emerging markets to explore innovation options for Nokia," according to its press release. Innovent also worked with my team. Both efforts kept Nokia well informed—and connected—to emerging disruptions. Nokia seemed to understand the landscape: from 1996 until 2007, it released a handful of mobile devices from its Communicator series, early versions of smartphones.

The company didn't continue these efforts. Innovent shut down. Collaboration with my group stopped in 2007 too. Nokia started to discount future and change. Nokia's market success started to dwindle shortly thereafter. Nowadays Nokia does not own the mobile phone division anymore, which at one point was synonymous with the company in most people's mind (Microsoft owned this part of Nokia's former operations starting in 2014 but shut it down in December 2017). Apple entered the smartphone business in 2007. Google introduced the Android operating system in 2008. And Samsung started building phones on that operating system soon after.

Nokia's timing to stop scanning efforts foreshadowed its demise. Relevant information couldn't reach decision-makers anymore; the company preoccupied itself with quantitative metrics. Nokia was not alone in complacently misjudging the industry. In 2012, Robert Safian, editor and managing director of *Fast Company*, summarized the development of the previous half decade. His statement is a lesson in the dangers of complacency: "Just five years ago, three companies controlled 64% of the smartphone market: Nokia, Research in Motion [better known as the maker of BlackBerry devices], and Motorola. Today, two different companies are at the top of the industry: Samsung and Apple. This sudden complete swap in the pecking order of a global multibillion-dollar industry is unprecedented."[31] Taking your eyes off the ball is deadly. You blink and the world moves on.

Christensen's 1997 book *The Innovator's Dilemma* compellingly outlines how companies can overemphasize current needs. Dedicated focus on what made them great also makes them fall when times move on. Kodak had become successful mainly with chemical innovations, and its clients still focused on chemicals-based photography—even at a time when digital photography already had established a very clear sales-growth trend line. Kodak never fully adapted to this change and became a prime example of the concept of an innovator's dilemma.

It doesn't have to be this way. Business strategists Larry Downes and Paul Nunes contrast Kodak's failure to adapt with Fuji Film's success. They show how Fuji Film leveraged its competency in chemicals and a portfolio of related capabilities to survive the transition to digital photography by pivoting successfully into the market of nanotechnology, flat-panel displays, and even cosmetics.[32] Here is another lesson to be learned. While Kodak had developed the first digital camera, Fuji Film not only foresaw a switch to digital technology as early as the 1980s but also developed a strategy that accounted for this switch.

The list of companies goes on; decision-makers become aware of disruptive powers. Innovation researchers Scott Anthony and Clayton Christensen argue, "Today's tightly interconnected markets make it harder for a company to be deaf to the roar of change."[33] Decision-makers still fail to take identification of disruptions seriously though. The question then is: Do you want to be the next Nokia, the next Kodak?

Abandoning Today in a Nutshell

Scanning provides market awareness of and for the future. But scanning can face insurmountable hurdles in organizations that neglect to prepare their operations for the future.

....

Short-term focus is not irrational behavior; short-term focus benefits many stakeholders. For organizations, short-term focus results in strategic myopia though. Interests are conflicting.

....

Quantitative obsession has increased with big data. Big data analytics comforts decision-makers in leaving machines in charge. But there is no replacement for human intelligence and intuition.

....

Finding the right *organizational home* is a full-blown effort. Don't skimp here. Scanning's impact, staff's perceived importance, and a functioning communication depend on the location.

....

Corporate culture is your friend and foe in embracing uncertainties or building hurdles to the future. Awareness is a competitive advantage. Corporate culture can favor complacency.

....

Hurdles are rarely signs of open rejection of or distrust toward scanning. Hurdles keep the status quo in place. Be aware of the hurdles; prepare for a struggle if you encounter them.

....

CHAPTER TEN

Inhabit Tomorrow

LEVERAGING YOUR NEWFOUND POWER

—◼—

Chance favors the prepared mind.

—Louis Pasteur

Big names can fail spectacularly when missing the world's turns. Nokia's utter loss of market dominance within a half decade was discussed previously. Kodak's fall from grace took slightly longer but was more complete. Kodak was among the five world's greatest brands in 1996. A dozen years on, the brand dropped from the list of the one hundred best global brands.[1] A brand associated with life's highlights—"Kodak Moments"!—vanished from public relevance essentially within a decade.

In contrast Apple and Amazon chose a different approach—the approach of scanning the world they want to inhabit. Apple did not create the graphical user interface; the claim goes to the Stanford Research Institute or Xerox Park (depending on the definition of such an interface). It did not develop music file sharing; Napster catapulted the application into public consciousness. It did not invent the smartphone; Nokia, of all companies, did that a decade earlier. And Amazon did not invent mail order nor cloud computing nor virtual assistants. Both companies—or rather their leaders—understood the importance of observation and preparation.

You cannot clone Steve Jobs or Jeff Bezos. But you can use scanning to capture their forward-looking capability. They learned to anticipate and prepare for market developments. Anyone can learn to do it. It just takes awareness, understanding, and discipline. And a tool like scanning can support you on the way.

FIPI provides a way to share knowledge and strategies among departments and with collaborating partners. It keeps all eyes looking forward. It facilitates talent development. It reinforces an agile and innovative culture. It gets results.

The Future in Reach

You overcame internal hurdles. You have a team that is eager to live in the future. You know how to turn seemingly random data points into convincing tales from the future. Now you can make sense of the future. If you are like most firms that look into the future, you will want to see where new opportunities await your arrival. You will, almost accidentally, discover many of the threats you have not considered. As you listen to tales from a world some years ahead, your mind will look at how your organization will or could fare in such a world. You will reevaluate your strengths and weaknesses in a world in which many assumptions became collateral damage and many distant plausibilities turned into realities now staring you in the face.

You have taken a mental tally of what the future might bring. You have run through a SWOT analysis: you investigated corporate strengths, weaknesses, opportunities, and threats. Opportunities and threats very obviously have to go beyond the current aspects of the marketplace, the here and now, to make any sense for strategic planning. With narratives about emerging worlds and remaining uncertainties, the outside world moves into focus. Now you can look at your own operations, see the strengths and weaknesses you have, and find the strengths and weaknesses you will need to identify.

Or you can use other tools you're familiar with. Scanning can move you forward in time to make better use of the five forces analysis that Michael Porter at the Harvard Business School introduced. The analysis looks at the rivalry within an industry, the bargaining power of buyers and suppliers, and the threats that emerge from new entrants or substitute products or services—the five forces. Since scanning offers a look at emerging dynamics within and across industries, your analysis will feature a forward-looking perspective that synchronizes your decision making with emerging market realities.

This exercise is what most companies set out to do when looking at scanning. Here is the point where many companies see scanning's promise fulfilled. But if that is all you use the process for, you are missing out on a bunch of benefits that are now as close as they can be, since you have already done most of the work. By better understanding the vagaries of the future, you opened a box of tools your team can work with. Patterns of events and developments as well as resulting narratives create anchor points for strategic thoughts and enable decision-makers to more effectively discuss emerging issues, exchange concerns, and take a preliminary look at potential strategic avenues to move the organization into a better position. Such capabilities translate across almost all cogs and wheels of organizational behavior.

Figure 15 highlights the four major areas of benefits scanning offers. Many of them fall into separate opportunities about how to leverage pattern creation of today's events to create narratives from the future. All of these benefits should be considered; all of them I have seen change the understanding of teams and individuals. Looking more broadly—and intensely—at the future can take the fear out of uncertainties and the sting out of surprises.

Scanning enables strategists to develop a laundry list of topics to respond to, monitor, or watch. Decision-makers then can juxtapose external signposts and internal milestones. *Signposts* are markers that indicate that a potential development is moving closer to becoming a reality. Your narrative of the future is turning into an account of the future. For instance, the first products that included graphene marked the point of commercialization. The first use of quantum computing outside of labs will make benefits and dangers actualities that will have to be considered in strategic plans going forward. Signposts structure narratives of the world into timelines. *Milestones* are the corporate capabilities and responses you throw at the market. Strategically, milestones relate to a number of alternatives that can move a company into a desirable market position. Signposts turn awareness into strategic necessities; milestones represent degrees of organizational preparedness.

The main purpose of scanning activities is to generate early strategic awareness of emerging and evolving developments—the foundation on which decision-makers can prepare their organizations for a changing future.

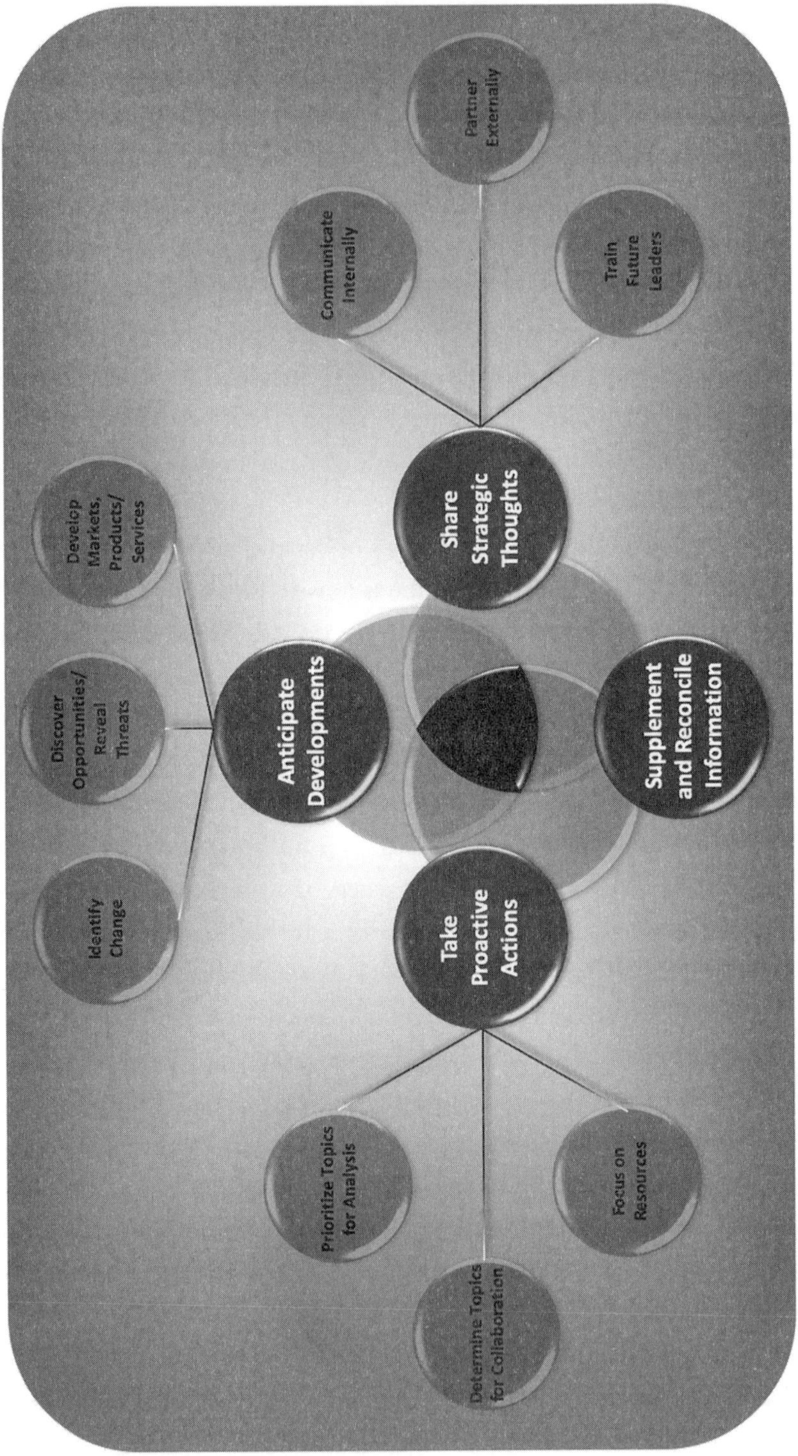

Figure 15: Benefits of Scanning

Scanning is a chance to bridge the awareness gap: don't miss the moment when you can be ahead of the curve. But certainly make sure that you won't fall behind. Bridge the gap between "Today's capacity for peripheral vision" and "Tomorrow's need for peripheral vision."[2] Use that bridge wherever you can. Scanning is flexible. Scanning is instructive. Scanning takes your teams out of their routines. Scanning constitutes a valuable tool in anybody's hands—if they are allowed to get their hands on it.

Anticipate—Don't Predict—the Future

Anticipation provides competitive advantages. Anticipation makes uncertainties manageable. Anticipation puts you in the lead. There are many ways to leverage memos you received from the future. Don't miss the power that comes with such early alerts.

Identify change. Change is like a box of chocolates; you never know what you're gonna get. But you can take the surprise out of it. Just knowing what might await you drastically confines uncertainties; suddenly, many potential surprises become expectable plausibilities. Just seeing what could be ahead of you puts you in a pole position no matter which alternative becomes reality. Change is inevitable, surprises are not.

But excitement about shifts can outrun market realities. Trends can prove short-lived. Fringe behavior might stay, well, fringe behavior. Other developments will gain momentum to become a growing market opportunity—in time. The maker movement has clear rationale and value. Makers not only know the problems they are trying to solve but also work on the solutions. It was easy to expect too much too early. But a look at how consumers try to address their needs themselves can be instructive. Author William Gibson found "the street finds its own uses for things"—therefore, you should look at the "street" to identify new uses and solutions.

Expectations can be disappointed, developments can be delayed, dynamics can shift. Many 3D printing companies expected that the maker movement would snap up their products to create cool contraptions and objects. At the height of the excitement, beginning in 2014, for instance, 3D printer providers Stratasys and 3D Systems saw their stocks reach above $135 and above $95, respectively. The following market disillusionment is reflected in the companies' valuation of the stocks at about $22 and $13, respectively, just two years later—while the technology-focused NASDAQ index gained substantially during the same period. (Since then the value of the 3D printer companies' stocks went to about $14 and $6 at the end of summer 2020 while the index hit an all-time high.) "We tend to overestimate the effect of a technology in the short run and underestimate the effect in the long run." And financial markets have an even shorter horizon than other markets.

New business models make assessments even more difficult. The dot-com boom and bust of the late 1990s until the beginning of 2000 highlights how an entire industry, arguably global economy, can get carried away. Early identification allowed priming the thought process, though. And preparation during that time period was the difference of "make it or break it" in the decade that followed. With hindsight, it appears that the bust was the necessary shakeout to pave the way for the success of Amazon, Facebook, Google, and hundreds of startups that since then have made names for themselves.

Scanning also serves practitioners well in anticipating social progressions. The increase in mobile applications very readily translates into commercial implications. The emerging Internet of Things will shape industries. Big data and AI are already here and will continue to change processes. Social developments are much less clear in scale, timing, and impact. But social developments permeate the entire fabric of societal and commercial life. They leave no industry, no commercial activity, unaffected. But they are very difficult to assess, they are less tangible, and they usually don't afford very clear implications and strategic direction. Women's rights and civil rights movements have been around for over a century even in the most conservative judgment, but arguably related commercial impact happened only in the 1960s. Now both movements have received new energy as the Me Too movement, Black Lives Matter, and ongoing gains of the LGBTQA community attest to. Social milestones are difficult to pin down on a timeline but pivotal to consider.

Anticipation is crucial. Anticipation is the foundation for competitive superiority. Scanning offers a way to look at today's world to see the future closing in. Day and Schoemaker spell out the payoffs: "Superior peripheral vision can help you anticipate risks and see opportunities sooner while gaining a profitable advantage over your blinkered rivals."[3]

Discover opportunities, reveal threats. Anticipation uncovers dynamics. These dynamics should not go unused. Anticipation is not the end goal. It's the beginning of your strategic exploration. Most naturally, a more detailed understanding of emerging opportunities and threats should plug into a tool that decision-makers have used for half a century. The SWOT analysis specifically connects a company's abilities—its strengths and weaknesses—with developments beyond its influence—the environment—to create opportunities and threats. Often such analysis misses a crucial part: the environment. Connections are made directly, such as "if we already produce plastic covers, we have an opportunity to produce plastic packaging too—a huge market." Perhaps you cringe. On the surface, the rationale makes sense. There's an existing manufacturing strength. There's a huge market that seemingly provides opportunities. Trendlines over the past decade show massive growth in plastic packaging; in fact, growth likely will continue in coming years. What can go wrong? For one, the world's attitude toward plastic is changing. Second, entering the market today could come with large liability issues tomorrow. Worth it? You be the judge. The example is obvious. The lessons should be clear though: anticipate the future to improve your existing tools. Scanning can elevate SWOT analysis from an assessment of the current situation to an evaluation of how emerging dynamics shift the analysis over time. Scanning helps analyze a strategy's future value.

Traditional SWOT analyses can be misleading. It is easy to forget the time needed to implement and execute a strategy that builds on today's—by then yesterday's—data (the growth line of plastic use). By the time your strategy is in place, the world has moved underneath your feet. Syncing today's analysis with tomorrow's world is the trick. Remember: "A great hockey player plays where the puck is going to be."

Peeking into the future also pinpoints emerging competitors—including from different industries using different technologies. Decision-makers are aware of their main competitors—a bread-and-butter necessity. Therefore, monitoring the usual suspects becomes second nature. The process of identifying and evaluating the unusual suspects is different. New startups, novel technologies, emerging business models, and changes in the value chain—what have you?—are outside of companies' purview. Out of sight, out of mind. Until you get punched in the mouth.

Scanning can be a very effective tool to identify threats and to anticipate risks. The vast majority of companies I have had the opportunity to work with focus on opportunities; new business is exciting. Protecting your business can be a slog; it's the grind of day-to-day work. Scanning can be a formidable platform to recognize threats, risks, and challenges at an early stage. Day-to-day work can become a forward-looking effort. In fact, a company I worked with ran into a testing market situation during the onset of the Great Recession. The company not only went through bankruptcy protection but also received financial help from the government. In return, the government wanted reassurances that the company would pay increased attention to commercial risks and market threats to ensure that the bailout wasn't wasted. The company decided to use scanning as one of the tools. Makes sense!

Develop markets and products or services. Anticipation of developments should only be the means to an end. Clearly, there must be emerging markets and needs that can be addressed. Anticipating the future gives us a chance to run ahead and prepare for the arrival of commercial opportunities.

Again, there's a well-known tool—an arrow that's already in most decision-makers' quiver. The Ansoff Matrix helps users sort through ideas to find expansion potential. Igor Ansoff developed a matrix that plotted existing and new markets (to the company) on one axis and existing and new products (and services) on the other axis. The matrix then delineates four quadrants that can provide corporations with growth alternatives or options for strategic shifts. Market penetration, market development, product development, and diversification provide very clear alternative paths to develop a company's fortunes. Again, most often the tool is used with information from the here and now. Again, strategies will run late when they hit the market. And again, scanning can sync timelines of strategy development and implementations with market developments.

Anticipating developments—identifying change, discovering opportunities and revealing threats, and *developing markets and products or services*—is at the core of most scanners' interest. Scanning opens a view into future worlds. Future worlds are where companies will have to move—hopefully early.

Such anticipation of developments is a crucial building block in developing mental future worlds to explore where to go and how to navigate hurdles and stumbling blocks. Building such future worlds is what scenario planning does. Scenario planning is the crucial tool for making strategic decisions under uncertainty; well, increasingly uncertainty is the status quo. Plus, scanning maps emerging developments, spells out uncertainties, and focuses on embracing these uncertainties.

Indeed, the US National Intelligence Council used my team's scanning efforts to identify trends that could shape the global future. *Global Trends 2025: A Transformed World* "uses scenarios to illustrate some of the many ways in which the drivers examined in the study . . . may interact to generate challenges and opportunities for future decision-makers."[4] My team also provided information for the follow-up discussions of future worlds. *Global Trends 2030: Alternative Worlds* highlighted complexity of global developments as a reason for using scenarios: "As our appreciation of the diversity and complexity of various factors has grown, we have increased our attention to scenarios or alternative worlds we might face."[5] The most recent publication, *Global Trends: Paradox of Progress*, updates related discussions.[6] Scenarios are narratives of plausible futures. Scanning provides the building blocks for such narratives.

Getting on the Same Page—Share the News

As in every organization, value accrues from making information available. Sharing strategic thoughts is key. The hurdle of failing to find a good strategic home will prevent anticipation of the future. But being in the wrong corner of an organization will prevent access to communication channels. What is the value of the best information if nobody knows about it? Sharing is crucial—advice that cannot be overstressed. The UK government concurs in its Horizon Scanning Programme review, underscoring the benefits of getting government organizations on the same page by establishing "shared views about the nature of plausible futures."[7]

Within an organization, silos prevent discussions; fiefdoms inhibit the exchange of ideas. Sharing of thoughts is imperative—to develop anticipation of the future and then to enable preparation of the organization. Max Bazerman and management scholar Dolly Chugh make the point, "Team members frequently discuss the information that they are all aware of, and they typically fail to share unique information with one another. Why? Because it's much easier to discuss common information and because common information is more positively rewarded as others chime in with their support."[8] Uncommon, new, even inconvenient information though is what scanning brings to the forefront. The information organizational members didn't expect; the information they don't want to hear—in short, the information they need to move forward.

Large corporations, in fact big divisions themselves, can resemble an assembly of silos. (Countries face the same challenges. The United Kingdom, for example has a Futures, Foresight and Horizon Scanning Programme in place.[9] In the US, futurist Amy Webb is calling for a Strategic Foresight Office, explaining that "there is no entity charged to focus on strategic foresight across domains equipped with the resources to undertake a comprehensive approach."[10]) Often these parts of the machinery work against each other; strategic partners hinder each other's efforts. No malicious intent needs to exist; failure finds its way very naturally. Information communication can be challenging, alignment next to impossible. Late Apple founder Steve Jobs was very aware of such challenges. The headquarters of Pixar, the animation film studio that he cofounded, were deliberately designed in a way to cause accidental run-ins of employees to foster socialization and to break down silos. "Ultimately, Jobs wanted the office layout to encourage unplanned meetings."[11] I'm not suggesting that you tear your buildings apart—although that might not be the worst idea—but I suggest paying deliberate attention to sharing of strategic thoughts.

The requirement to get information to people who need it is not a surprise; how to do it can be a challenge though. Sharing of information is the second-most commonly stated reason for employing scanning as a process. But I still encounter situations in which teams in organizations work in parallel on the same problem, being unaware of each other's work thereby duplicating efforts. Or departments pursue opposing goals, negating each other's efforts. It is difficult to decide which of these situations is more problematic. Collaboration across value chain partners can be even more daunting: intrinsic conflicts of interest exist. Discussions in workshops allow you to scratch the surface of the marketplace's veneer; narratives are fabulous shortcuts to understanding. "Use dialog to share the big picture,"[12] Day and Schoemaker advise. Such dialog can occur on many levels.

Communicate internally across and between functions and operations. Invite members of different functional teams to participate in the process; make them evangelists of scanning. Different functions hold varied knowledge and therefore can contribute multiple perspectives and insights. R&D can contribute technological information, marketing can provide insights into emerging consumer behavior, and logistics can offer information on new methods of operational efficiencies. That knowledge enriches discussions; breadth of considerations is the crucial input. Varied perspectives matter. As the saying goes, "every threat is also an opportunity." Now the opposite holds true too—unfortunately.

Understanding different perspectives helps you investigate developments. In the summer of 2015, package-delivery firm United Parcel Service inaugurated a test facility to look at the use and potential of 3D printing technology—the additive-manufacturing approach that allows printing of products from raw materials, often plastics, to form three-dimensional objects. (Remember, this was the time of general excitement about the technology's applications.) The company wanted to see whether the technology would threaten its operations or could be a valuable enabler. The question was if 3D printing would reduce shipments as on-location manufacturing becomes common. Or could the technology establish a chance for UPS to enter the local, on-demand production business? As mentioned before, much of the initial 3D printing hype has evaporated. But underestimating technologies' effect in the long run is the norm. Develop understanding early; reduce surprises!

Internal communication can bring information together. In one of the Scan meetings I moderated in the middle of the first decade of the new millennium, a Japanese company had sent a number of participants from a wide range of functions. The purpose was to enhance communication. One of the discussions revolved around new technologies to automatically capture speech, analyze it, and extract information. Note that the introduction of Siri, the speech-technology-enabled personal assistant, in 2011 was still about a half decade away. At the time, the technology was still considered speculative outside of dedicated R&D labs and some defined tasks. The representative from the marketing department mentioned how helpful such a function would be to analyze customer reactions and feedback on the phone. Representatives from the R&D department then voiced surprise given that they had developed a prototype a year earlier. They shelved the project though because they couldn't find a promising application. The coincidence is fortuitous. Don't expect such an outcome from every session. Illustrative of such sessions' power it is nevertheless.

Scanning can create a more collaborative mindset, an environment to look into the future. Establishing new communication channels and strengthening existing ones are side effects of trying to involve diverse teams in scanning. In 2012 I had an opportunity to moderate a scanning workshop at an annual meeting of a German chemical company. (The meeting took place outside of San Juan on the island of Puerto Rico. This way the company not only was able to attract international employees from a wide range of functions but also established an atmosphere to avoid a business-as-usual mindset.) The scanning event provided participants relief from chores more commonly associated with presentations and workgroups at such events. More important, representatives from various corners of the firm were able to interact. Such interactions establish ties that continue after such events.

Scanning is about weak signals. Weak signals—per definition—are difficult to sense. Communicating them can be an uphill battle too. First, the strong signals of current fires drown out the weak signals from the future. Second, how many people in an organization think and breathe weak signals? The burden can be too large to push through a potential "if" when current "ares" bind the corporations' attention. Participation in workshops and communication of narratives can overcome some of the hurdles.

Partner externally with outsiders across the value chain and industries. Communication within companies is challenging. Now try the same across value or supply chain partners! Organizational structures can be very different, and goals are diverging. The distance to the ultimate market, often the end consumer, varies across the supply chain. Information gets lost along the chain of partners—not unsimilar to the telephone game in which children pass on a message by whispering into the next child's ear; the final message received can be very different from the initial one sent. Scanning can get partners on the same page—at least to read from the same book. I repeatedly have been involved in workshops in which chemical companies collaborated with their customers to better understand their needs. By listening to the concerns and issues their customers voiced, participants from the supplying chemical company can start to anticipate future needs.

Scanning can articulate tomorrow's needs. Working with a telecommunications-equipment provider, I engaged through scanning workshops with city authorities in Boston, Massachusetts; Chicago, Illinois; and Copenhagen, Denmark. While I was delivering my narratives from past scanning workshops, the listening representatives of various authorities were able to envision capabilities and challenges of the future. The customers' voices became loud and clear to the equipment firm. Scanning can be a mouthpiece.

Scanning can elucidate uncertainties. It can build narratives for things you know that you don't know enough about. (Well, nobody really knows enough about things concerning the future.) First, some developments envelop multiple industries simultaneously, and entire economies can go through rapid changes. Frictions and pressures can be palpable but dynamics unclear, outcomes even less so. Second, other developments are so amorphous, shapeless, and undefined that they require an outside view. What do other industries think? What does it mean for them, if anything?

In 2010, I had the opportunity to work with Kraft Foods Inc. The company decided to engage in an open meeting with a wide range of partners and industry outsiders. The company invited representatives from many industries, including the agricultural-equipment, automotive, consumer packaged goods, electronics-components, food-processing, home-appliances, industrial-manufacturing, medical-device, and telecommunication-equipment industries. Discussions of the future quickly gravitated toward two areas of interest—issues that affected them all. Developments in global talent availability and ways to attract promising talent became clear concerns. The other topic that all participants wanted to understand was the at-the-time emerging mobile marketplace and its potential implications. Both topics still resonate with most industries—likely more today than at the time.

Also in 2010, a semiconductor company invited a wide range of companies to understand perspectives from the automotive, defense, telecommunication-equipment, trade, and software industries. Participants discussed the emerging concept of "personal digital persona." The purpose was to develop a definition of the concept and future implications. The goal was to understand what such a persona might mean in the context of different industries and applications. In contrast to the meeting at Kraft, participants in this workshop had very distinct views about what such a persona could be and what implications might develop. Whereas some industries had a very indifferent outlook, the defense industry in particular showed concern about issues related to cybersecurity. Other industries saw the concept more positively: marketing and communication opportunities could emerge. Personal digital persona meant many different things to different industries; its relevance ranged from virtually nonexistent to potentially game changing. Nevertheless, all participants left with a better understanding of what the future might bring.

Train future leaders. Peeking into the future allows future managers to train their senses to tune into potential changes to then consider appropriate strategies. Young talent tends to be educated and trained in one functional discipline (and often within that discipline in very specific aspects). The world has become increasingly specialized. Companies now face the challenge of using such expertise in a broader context. Scanning provides a sense of importance of interdisciplinary efforts. It trains us to keep an eye on the broader context. It allows us to see pathways to the future.

In my experience, Japanese companies in particular tend to see the benefits of employing scanning as a training tool. I have been involved in setting up and running scanning meetings in this region with the explicit goal of educating junior managers on the relevance of understanding dynamics beyond the realm of the company more often than in other regions. Perhaps the feeling of two lost decades drives this urgency. In the end, it does not matter whether this use is of particular regional interest; experience shows that scanning can be a very effective tool in training staff and future leaders.

Getting younger talent involved also provides a different set of perspectives and interests. And participants trained in business will learn to appreciate how advances in technology can dramatically affect business models. Similarly, scientifically educated team members will learn to appreciate that the best technology still requires a commercial breeding ground to succeed in the market. Decision-makers will understand how pieces fit together. A comprehensive picture will emerge. Brushstrokes create an image. Tiles form a mosaic.

Sharing information and developing an understanding of how the future might operate are crucial requirements. Speaking at the Forum on Leadership at the George W. Bush Presidential Center in April 2018, Amazon's CEO Jeff Bezos emphasized, "You do need the data, but then you need to check that data with your intuition and your instincts. And you need to teach that to all the senior executives and junior executives."[13]

Check Your Assumptions

Scanning is never the only tool you should use. It's part of a portfolio of tools at your disposal to make sense of the world. It can be a great arbiter of assumptions though. It supplements and reconciles information you already have. It can confirm instincts; it can challenge assumptions. It can round out your perspective on things to come.

Internal market research, business intelligence, operational analytics, and recently the addition of big-data-based approaches can swamp companies with information—usually about the past, sometimes about the now. Lagging indicators such as past quarters' growth and sales dominate many businesses' information. Trends show what is happening today. Tomorrow's world usually finds representation in predictions—predictions often made only to justify investments—invariably wrong. Most companies employ a mix of all these information sources. Organizational implementation can be arbitrary, even haphazard.

I am not arguing that scanning is the crucial tool in this portfolio because it is more important than other methods and analyses. I argue that it is so crucial because it tends to be overlooked. Therefore, it can provide some missing pieces. It can put other information in context. It's the approach that truly tries to coax emerging developments out of the sea of data. It's the one tool that solely looks at future dynamics.

Supplementing information from other sources should not be an afterthought. While other information tends to be available to all market participants, scanning provides you with a truly unique perspective. Common information sources reinforce existing views. They develop information bubbles. Existing views need to be challenged. Information bubbles require being burst.

Looking at different points of view is a worthwhile exercise. Once you embark on your strategy, you might put your company on tracks to nowhere. If scanning confirms your intuition and insights, no harm is done. You will better understand the environment you're stepping into. You will see red flags earlier. If scanning results in a very different narrative of the future than the one you currently hold, you should pause. Perhaps the tracks you intend to take are dead-end roads.

Take Steps toward the Future

If you engage in the scanning process without taking first steps toward the future, you're indulging in corporate entertainment. Likely worse. Your team will feel prepared for the future, but in reality it has only come to realize what the future might look like. The crucial part is missing. All the effort you took has only one meaningful purpose: to take proactive actions to position yourself at the right place to let the future meet you there.

The final, and for all intents and purposes most important, component of scanning efforts therefore has to be the impact/emergence matrix, which segues from topic identification and understanding into *initiating follow-up steps to take proactive action*. A number of benefits exist.

Focus on resources. Resources are limited; there is no insight in this statement. There is also no mystery in the fact that leaders of different functions might see their own work as the pivotal ingredient for a corporation's success. First-hand experience with technology-driven innovation approaches through various collaborations in projects and initiatives shows how often, particularly in Silicon Valley, technology is perceived as the ends, not a means to an end. But technology serves the purpose to achieve commercial relevance.

SRI International is a research institute that is built on scientific and technological excellence. The organization can claim, among many other inventions and innovations, a pivotal role in developing ARPANET (which developed into the internet and World Wide Web), the development of the graphical user interface and the mouse, and Shakey, the first robot that had a rudimentary ability to reason about its actions. Imaging technologies, speech recognition (Apple's Siri emerged from SRI's laboratories), swarm robotics, stealth technologies, artificial intelligence, and medical research are all important parts of SRI International's work efforts. To say the institute is focused on technological advances perhaps is an understatement. But Curtis Carlson, who was SRI International's CEO for more than one and a half decades, was very aware that technological advances and innovation cannot become impactful when confined to a commercial vacuum. He acknowledges market relevance as the driver of success. He suggests, "Focus your attention on important rather than interesting needs" by asking "Does anyone care?"[14]

The impact/emergence matrix offers an early way to exclude developments that have no bearing on a corporation's commercial well-being. The matrix seamlessly segues from discussion to corporate responses, providing the glue that splices strategic awareness with organizational preparedness.

Determine topics for collaboration. Scanning attracts personnel from a wide range of functions and disciplines across the organization that receive an opportunity to exchange ideas, concerns, and opportunities. Understanding of other departments' challenges can quickly be developed and shared. Silos can be broken up. A better understanding of how your work relates to other developments emerges. Researcher Joshua Krook argues, "The more specialised our workforce becomes, the less capable we are of seeing how our industry relates to other industries."[15] The nature of scanning puts a premium on working together to identify and assess developments outside of a corporation.

At the Kraft Foods meeting, participants from about ten industries discussed events that might affect their future business. Two topics stood out: global talent and the mobile marketplace. The topics applied across company and industry boundaries. During the meeting, representatives from various companies started to collaborate on these topics and shared company-specific experiences. A number of company representatives then decided to stay in contact and work together on these topics. Cross-industry collaboration cannot be any more easily initiated.

The workshop on "digital persona" had a very different outcome. The idea of digital persona proved to have no common ground among participants. The discussion nevertheless highlighted a variety of concerns that had relevance for individual participants—quickly focusing their interest on the aspects most likely to affect their respective industries.

Prioritize topics for analysis. Scanning points to developments and changes. True analysis of factors and implications requires further research. Also, decision-makers have to figure out what the implications for the various parts of an organization are. The impact/emergence matrix aims at maintaining momentum to seamlessly segue to action steps by prioritizing identified developments. Not all changes affect an organization. Not all changes will require urgent action. But some changes justify in-depth analysis. Understanding the future requires work.

Because technological developments have proven particularly fast-paced and complex, many companies see scanning as a crucial input to identify and map emerging technologies. What it means for operations and markets requires studying. Further analysis also can determine the commercial readiness of technologies. Scanning should trigger further curiosity. Scanning does not allow for—and is not meant to provide—in-depth analysis and quantitative evaluation of technologies. This step should follow any broad look at the future.

Take your next steps to inhabit the future. Awareness is a prerequisite. Preparation is the success factor. After all, "Chance favors the prepared mind." Markets favor prepared companies.

Inhabiting Tomorrow in a Nutshell

Scanning provides market awareness of and for the future. Use such awareness conscientiously. Don't squander the market advantage you gained.

....

Anticipate developments so that you can be ahead of the curve. Now that you understand the potential for change, look at your opportunities and take threats seriously. Develop markets before they fully emerge.

....

Share strategic thoughts to bring team members and partners on the same page. Communicate novel insights generously within your organization; knowledge fiefdoms create insight vacuums. Work with your partners. And use the process as a training tool to create management potential.

....

Supplement and reconcile information to ensure quality and authenticity of insights. Treat confirmed insights as precious diamonds in the rough. Be concerned about contradicting information. Double-check and reiterate.

....

Take proactive actions to make your work count. Creating awareness without letting action follow is corporate entertainment. Your efforts then can become a distraction. Only actions count in the marketplace. Understanding without actions is academic at best.

....

Scanning enables preparation for the future. Anticipation is the foundation; preparation is the house you build. If you are prepared, surprises become expectations and market changes will be the ladder to success.

....

The Future Starts Now

Cyberpunk novelist William Gibson pronounced, "The future is already here—it's just not very evenly distributed." He's right. But redistributing the future to make sense of it takes more than good intentions and a lucky charm. It requires a methodology, a process, and a set of practical tools that can help you take your organization from its present position in the market to greater dominance in the future. Scanning is just such a tool.

The future is uncertain. Not looking at your left and at your right of what is happening is a surefire way to become strategically misguided, or in the worst case commercially obsolete. Some of the seemingly best run companies at the time failed to look at the big picture and to consider the full set of options and alternatives. Forecasts and predictions led them astray. The future is uncertain and will remain so.

The good news is that the ingredients of the future are already in place. A new day does not start from square one; many preceding steps already point the way. The task, though, is to identify early signs of changes that will shape the future. Identifying developments and dynamics that exist today and considering how they can conflict, coalesce, or reinforce each other can prepare decision-makers. The future is uncertain. Potential pathways to alternative futures do not have to be; they just require effort, time, and openness to identify and consider. "The future is already here—it's just not very evenly distributed."

Scanning is not, cannot be, the only tool decision-makers should employ. Similar to a toolbox, one single tool will not suffice to get the job done. But scanning is a tool that has been neglected by many companies and undervalued by others for too long. Scanning can guide your research. Scanning can offer the competitive advantage needed to succeed in an increasingly complex marketplace. Scanning forces decision-makers to consider a much wider range of possibilities than most companies are willing to consider.

In *The Empire Strikes Back* of the original *Star Wars* trilogy, fictional Jedi Master Yoda responds when asked about the future, "Difficult to see. Always in motion is the future." The future indeed cannot be foreseen, but the forces and dynamics that will shape the future can be identified.

The world will remain uncertain. But rather than dreading uncertainty, you can make uncertainty your ally in creating competitive advantage. After all, uncertainty affects us all. But the winners of the future know that if you do not take it into account, it will ambush you.

When I started my work with scanning, I had to abandon my preoccupation with today. Over the past two decades, tomorrow became today's work for me. It took a shift in thinking about marketplaces. Scanning is neither a magical crystal ball nor a Ouija board. It is, however, a simple step-by-step process that anyone can use to create a better future for an organization.

Where does today end? Yesterday. When does tomorrow begin? Now.

Notes

Chapter One

1 "Global Market Share Held by Nokia Smartphones from 1st Quarter 2007 to 2nd Quarter 2013."
2 Patton, "The Role of Scanning in Open Intelligence Systems."
3 Buchanan, "Power Laws & the New Science of Complexity Management."
4 Schoemaker with Gunther, *Profiting from Uncertainty.*
5 Smith and Kruzic, *Analyzing Future Business Environments.*
6 Schoemaker with Gunther, *Profiting from Uncertainty.*
7 Patton, "The Role of Scanning in Open Intelligence Systems."
8 Courtney, *20/20 Foresight.*
9 Schoemaker with Gunther, *Profiting from Uncertainty.*
10 *Handbook of Knowledge Society Foresight.*
11 *Handbook of Knowledge Society Foresight.*
12 Schoemaker with Gunther, *Profiting from Uncertainty.*
13 Rohrbeck, *Corporate Foresight.*
14 Madsbjerg and Rasmussen, "The Power of 'Thick' Data."
15 Taleb, "Beware the Big Errors of 'Big Data.'"
16 Martin, "Beyond the Numbers."
17 Toffler, *The Third Wave.*
18 Martin, "Beyond the Numbers."
19 Morson and Schapiro, *Cents and Sensibility.*

Chapter Two

1 Ralston and Wilson, *Scenario Planning Handbook.*
2 Tetlock and Gardner, *Superforecasting.*
3 *Handbook of Knowledge Society Foresight.*

4   Ramírez, Churchhouse, Palermo, and Hoffmann, "Using Scenario Planning to Reshape Strategy."

5   Patton, "The Role of Scanning in Open Intelligence Systems."

6   Aguilar, *Scanning the Business Environment*.

7   Marketing material, SRI International.

8   "A Dip into a Think Tank."

9   Drucker, "Looking Ahead: Implications of the Present."

10 McGrath, *Seeing around Corners*.

11 Sargut and McGrath, "Learning to Live with Complexity."

12 Mackintosh, "Three Economists Walk into a Bar."

13 Morson and Schapiro, *Cents and Sensibility*.

14 McGrath, "Is Your Company Ready to Operate as a Market?"

15 Thompson, "The Coming Crisis: We're Not in Kansas Any More."

16 Bazerman and Watkins, *Predictable Surprises*.

17 Tetlock and Gardner, *Superforecasting*.

18 Courtney, *20/20 Foresight*.

19 *Handbook of Knowledge Society Foresight*.

20 Kranzberg, "Technology and History: 'Kranzberg's Laws.'"

21 Kane, "Digital Disruption Is a People Problem."

22 "Fraunhofer Creates New Group for Innovation Research."

23 Day and Schoemaker, "Are You a 'Vigilant Leader'?"

Chapter Three

1   Rosling, *Factfulness*.

2   Gorbis, *The Nature of the Future*.

3   Novet, "Amazon Says AWS Revenue Jumped 46 Percent in Third Quarter."

4   Clay, *Electronic Delivery and Distribution of Music*.

5   Maskell, *Music Consumer Insight Report 2016*.

6   Watson, "Trading Places: Global Box Office Dethroned by Spending on Home Entertainment."

7   Sargut and McGrath, "Learning to Live with Complexity."

8   Day and Schoemaker, *Peripheral Vision*.

9   Bazerman and Watkins, *Predictable Surprises*.

10 Taleb, "Beware the Big Errors of 'Big Data.'"

11 Martin, "Beyond the Numbers."

12 Courtney, *20/20 Foresight*.

13 Day and Schoemaker, "Scanning the Periphery."

14 Saffo, "Six Rules for ~~Accurate~~ *Effective* Forecasting."

15 Alker, *Conversing with Images*.

16 Schwirn, *Leveraging Photography on the Web*.

Chapter Four

1 Aguilar, *Scanning the Business Environment*.

2 *Handbook of Knowledge Society Foresight*.

3 Schoemaker with Gunther, *Profiting from Uncertainty*.

4 Day and Schoemaker, *Peripheral Vision*.

5 Webb, *The Signals Are Talking*.

6 Christensen, *The Innovator's Dilemma*.

7 Christensen, Raynor, and McDonald, "What Is Disruptive Innovation?"

8 Christensen, Raynor, and McDonald, "What Is Disruptive Innovation?"

9 *Wikipedia*, online entry.

10 Bazerman and Watkins, *Predictable Surprises*.

11 Smith and Dumaine, "How I Delivered the Goods."

12 Anderson, "The Long Tail."

13 Aguilar, *Scanning the Business Environment*.

14 "Oxford Dictionaries Word of the Year 2016 Is. . . . "

15 Anderson, "The Long Tail."

16 Anderson, "The End of Theory."

17 Pentland, "Reality Mining."

18 Koppel, *Lights Out*.

19 "Alert (TA17-293A) Advanced Persistent Threat Activity Targeting Energy and Other Critical Infrastructure Sectors."

20 "Alert (TA18-074A) Russian Government Cyber Activity Targeting Energy and Other Critical Infrastructure Sectors."

21 Swift, "To Apple, Love Taylor."

22 Lindstrom, *Small Data*.

23 Hersh, "Closing Conversation with Jeff Bezos."

24 Grove, *Only the Paranoid Survive*.

Chapter Five

1 Broderick, *Automation Climbing the Value Chain.*
2 Brynjolfsson and McAfee, *Race Against the Machine.*
3 Pistrui, "The Future of Human Work Is Imagination, Creativity, and Strategy."
4 McGrath, *Seeing around Corners.*
5 Webb, *The Signals are Talking.*
6 Tighe, *Rethinking Strategy.*
7 Brynjolfsson and McAfee, *Race Against the Machine.*
8 O'Reilly, editorial review of *Race Against the Machine.*
9 Broderick, *Automation Climbing the Value Chain.*
10 Manyika, Chui, Bughin, Dobbs, Bisson, and Marrs, *Disruptive Technologies.*
11 Pistrui, "The Future of Human Work Is Imagination, Creativity, and Strategy."
12 Brynjolfsson and McAfee, *Race Against the Machine.*
13 Broderick, *Automation Climbing the Value Chain.*
14 Telford, *Robot Representation.*
15 Schwirn, *Autonomous Robots in the Wild.*
16 "Make No Mistake: To Err IS Human."
17 Beinhocker and Kaplan, "Tired of Strategic Planning?"
18 Dufva, *Knowledge Creation in Foresight.*
19 Dufva, *Knowledge Creation in Foresight.*
20 Gorbis, "Five Principles for Thinking Like a Futurist."
21 Dufva, *Knowledge Creation in Foresight.*
22 Ansoff and McDonnell, *Implanting Strategic Management.*
23 Grove, *Only the Paranoid Survive.*
24 Day and Schoemaker, "Scanning the Periphery."
25 Adner, *The Wide Lens.*
26 Mims, "Five Tech Predictions for the Year Ahead."
27 Schwirn, *URLs for Products: The Internet of Things.*
28 Schwirn, *A Different Kind of "Internet of Things."*
29 Manyika, Chui, Bughin, Dobbs, Bisson, and Marrs, *Disruptive Technologies.*
30 Hoffmann, *eScience.*
31 Telford, *Robot Representation.*
32 Schwirn, *Autonomous Robots in the Wild.*
33 Day and Schoemaker, "Scanning the Periphery."
34 Morson and Schapiro, *Cents and Sensibility.*

Chapter Six

1   Schwartz, *The Art of the Long View.*

2   Schwirn, *Automation on Top of the Value Chain.*

3   Schwirn, *Automation on Top of the Value Chain.*

4   Peng, *Self-Tracking Health Data.*

5   Schwirn, *Diagnosed Self.*

6   Schwirn, *Quantified Everybody.*

7   Schwirn, *Quantified Employees.*

8   Telford, *Quantified Communities.*

9   Harris, *Small Countries, Big Innovations.*

10 Pentland, "Reality Mining."

11 Patton, *Reality Mining and the DataTurbine.*

12 Hinzmann, *Next-Generation Consumers.*

13 Peng, *Community Values Trump Individualism.*

14 Wiesbrock, *Collaborative Consumption.*

15 Hof, "The Power of Us."

16 Schwirn, *From Cocreation to Competition.*

17 Schwirn, *Unexpected Competition from Smartphones.*

18 Hollister, "The Age of the iPod Is Over."

19 Paul, "Traditional GPS Is Dead. Long Live Smartphone GPS."

20 Zhang, "This Latest Camera Sales Chart Shows the Compact Camera Near Death."

21 Haines, "The E-Reader Device Is Dying a Rapid Death."

22 Wijman, "Mobile Revenues Account for More than 50% of the Global Games Market."

23 Ralston and Wilson, *Scenario Planning Handbook.*

24 Dufva, *Knowledge Creation in Foresight.*

25 Wilkinson and Kupers, "Living in the Futures."

26 Morson and Schapiro, *Cents and Sensibility.*

27 *FOREN Guide: Foresight for Regional Development Network.*

28 Madsbjerg, *Sensemaking.*

29 Blum, Goldfarb, and Lederman, "The Path to Prescription."

30 *Handbook of Knowledge Society Foresight.*

31 *Handbook of Knowledge Society Foresight.*

32 Schwartz, *The Art of the Long View.*

33 Bezos, "Six-Page Narratives."

34 Madsbjerg, *Sensemaking*.

Chapter Seven

1 *Handbook of Knowledge Society Foresight*.

2 Bazerman and Watkins, *Predictable Surprises*.

3 Saffo, "Six Rules for ~~Accurate~~ *Effective* Forecasting."

4 Gates, "A Better Response to the Next Pandemic."

5 Gates, "The Next Outbreak? We're Not Ready."

6 "Bill Gates: We Are Vulnerable to Flu Epidemic in Next Decade."

7 Zambon, *Global Health Experts Advise Advance Planning for Inevitable Pandemic*.

8 Morrison, "Biodefense Summit Transcript."

9 The Council of Economic Advisers, *Mitigating the Impact of Pandemic Influenza*.

10 Schwirn, *Life After the Time of Coronavirus*.

11 Morvan and Wei, *Infected*.

12 Morvan and Wei, *Infected*.

13 Samuels, "Depression Rate Has Tripled among US Adults."

14 Petrache, *Halt the Epidemics!*

Chapter Eight

1 Ansoff and McDonnell, *Implanting Strategic Management*.

2 Schoemaker with Gunther, *Profiting from Uncertainty*.

3 Ansoff and McDonnell, *Implanting Strategic Management*.

4 Day and Schoemaker, *Peripheral Vision*.

5 Weick and Sutcliffe, *Managing the Unexpected*.

Chapter Nine

1 Zetlin, "Blockbuster Could Have Bought Netflix."

2 Benzinga, "Remember When Yahoo Turned Down $1 Million to Buy Google?"

3 Bazerman and Watkins, *Predictable Surprises*.

4 "Horizon Scanning Programme: A New Approach for Policy Making."

5 Barton and Wiseman, "Focusing Capital on the Long Term."

6 Bazerman and Watkins, *Predictable Surprises.*

7 Bazerman, *The Power of Noticing.*

8 Day and Schoemaker, *Peripheral Vision.*

9 Tighe, *Rethinking Strategy.*

10 McGrath, *Seeing around Corners.*

11 Tighe, *Rethinking Strategy.*

12 Tetlock and Gardner, *Superforecasting.*

13 Anderson, "The End of Theory."

14 Coy, "Lies, Damn Lies, and Financial Statistics."

15 Burns and Novick, *The Vietnam War.*

16 Burns and Novick, *The Vietnam War.*

17 Rosling, *Factfulness.*

18 Burns and Novick, *The Vietnam War.*

19 Harford, "Big Data: Are We Making a Big Mistake?"

20 Bhidé, "The Big Idea: The Judgment Deficit."

21 Silver, *The Signal and the Noise.*

22 Taleb, "Beware the Big Errors of 'Big Data.'"

23 Martin, "Beyond the Numbers."

24 Day and Schoemaker, *Peripheral Vision.*

25 Aguilar, *Scanning the Business Environment.*

26 Ansoff and McDonnell, *Implanting Strategic Management.*

27 Day and Schoemaker, "Scanning the Periphery."

28 Bazerman and Watkins, *Predictable Surprises.*

29 Weick and Sutcliffe, *Managing the Unexpected.*

30 Bazerman and Watkins, *Predictable Surprises.*

31 Safian, "This Is Generation Flux."

32 Downes and Nunes, "Big-Bang Disruption."

33 Anthony and Christensen, "The Empire Strikes Back."

Chapter Ten

1 *The World's Greatest Brands.*

2 Day and Schoemaker, *Peripheral Vision.*

3 Day and Schoemaker, *Peripheral Vision.*

4 *Global Trends 2025: A Transformed World.*

5 *Global Trends 2030: Alternative Worlds.*

6 *Global Trends: Paradox of Progress.*

7 "Horizon Scanning Programme: A New Approach for Policy Making."

8 Bazerman and Chugh, "Decisions Without Blinders."

9 "Futures, Foresight and Horizon Scanning."

10 Webb, "A National Office for Strategic Foresight."

11 D'Onfro, "Steve Jobs Had a Crazy Idea."

12 Day and Schoemaker, *Peripheral Vision.*

13 Hersh, "Closing Conversation with Jeff Bezos."

14 Carlson, *Innovation.*

15 Krook, "Why Experts Have Killed Innovation."

Bibliography

Adner, Ron, *The Wide Lens: What Successful Innovators See That Others Miss*, 2013.

Aguilar, Francis J., *Scanning the Business Environment*, 1967.

"Alert (TA17-293A) Advanced Persistent Threat Activity Targeting Energy and Other Critical Infrastructure Sectors," US-Cert (United States Computer Emergency Readiness Team), result of analytic efforts between the Department of Homeland Security (DHS) and the Federal Bureau of Investigation (FBI), revised release October 23, 2017.

"Alert (TA18-074A) Russian Government Cyber Activity Targeting Energy and Other Critical Infrastructure Sectors," United States Computer Emergency Readiness Team (US-CERT), March 15, 2018.

Alker, Dave, *Conversing with Images*, SRI Consulting Business Intelligence, February 2004.

Anderson, Chris, "The End of Theory: The Data Deluge Makes the Scientific Method Obsolete," *Wired*, June 23, 2008.

Anderson, Chris, "The Long Tail," *Wired*, October 1, 2004.

Ansoff, Igor, and Edward McDonnell, *Implanting Strategic Management*, second edition, 1990.

Anthony, Scott D., and Clayton M. Christensen, "The Empire Strikes Back," *MIT Technology Review*, January/February 2012.

Barton, Dominic, and Mark Wiseman, "Focusing Capital on the Long Term," *Harvard Business Review*, January–February 2014.

Bazerman, Max H., *The Power of Noticing: What the Best Leaders See*, 2014.

Bazerman, Max H. and Dolly Chugh, "Decisions without Blinders," *Harvard Business Review*, January 2006.

Bazerman, Max H., and Michael D. Watkins, *Predictable Surprises: The Disasters You Should Have Seen Coming and How to Prevent Them*, 2008.

Beinhocker, Eric D., and Sarah Kaplan, "Tired of Strategic Planning?," *McKinsey Quarterly*, June 2002.

Best Global Brands, Interbrand, 2008.

Bezos, Jeff, "Six-Page Narratives," *2017 Letter to Shareholders*, April 18, 2018.

Bhidé, Amar, "The Big Idea: The Judgment Deficit," *Harvard Business Review*, September 2010.

"Bill Gates: We Are Vulnerable to Flu Epidemic in Next Decade," *BBC News*, December 30, 2016.

Blum, Bernardo, Avi Goldfarb, and Mara Lederman, "The Path to Prescription: Closing the Gap between the Promise and the Reality of Big Data," *Harvard Business Review*, September 2015.

Broderick, Andrew, *Automation Climbing the Value Chain*, SRI Consulting Business Intelligence, November 2006.

Brynjolfsson, Erik, and Andy McAfee, *Race Against the Machine: How the Digital Revolution Is Accelerating Innovation, Driving Productivity, and Irreversibly Transforming Employment and the Economy*, 2011.

Buchanan, Mark, "Power Laws & the New Science of Complexity Management," *strategy+business*, Spring 2004.

Burns, Ken, and Lynn Novick, *The Vietnam War* (television documentary series), explicit language version, 2017.

Carlson, Curtis, *Innovation: The Five Disciplines for Creating What Customers Want*, 2006.

Christensen, Clayton, *The Innovator's Dilemma: When New Technologies Cause Great Firms to Fail*, 1997.

Christensen, Clayton, Michael E. Raynor, and Rory McDonald, "What Is Disruptive Innovation?," *Harvard Business Review*, December 2015.

Clay, Molly Ruffin, *Electronic Delivery and Distribution of Music*, SRI Consulting, April 1998.

The Council of Economic Advisers, *Mitigating the Impact of Pandemic Influenza through Vaccine Innovation*, September 2019.

Courtney, Hugh, *20/20 Foresight: Crafting Strategy in an Uncertain World*, 2001.

Coy, Peter, "Lies, Damn Lies, and Financial Statistics," *Bloomberg Businessweek*, April 6, 2017.

Day, George S., and Paul J. H. Schoemaker, "Are You a 'Vigilant Leader'?," *MIT Sloan Management Review*, April 1, 2008.

Day, George S., and Paul J. H. Schoemaker, *Peripheral Vision: Detecting the Weak Signals That Will Make or Break Your Company*, 2006.

Day, George, and Paul J. H. Schoemaker, "Scanning the Periphery," *Harvard Business Review*, November 2005.

Derrick, Jayson, "Remember When Yahoo Turned Down $1 Million To Buy Google?," Yahoo! Finance, July 25, 2016.

"A Dip into a Think Tank," *Time*, November 30, 1981.

D'Onfro, Jillian, "Steve Jobs Had a Crazy Idea for Pixar's Office to Force People to Talk More," *Business Insider*, March 20, 2015.

Downes, Larry, and Paul Nunes, "Big-Bang Disruption," *Harvard Business Review*, March 2013.

Drucker, Peter, "Looking Ahead: Implications of the Present," *Harvard Business Review*, September/October 1997.

Dufva, Mikko, *Knowledge Creation in Foresight: A Practice- and Systems-Oriented View*, Aalto University publication series, January 2016.

FOREN Guide: Foresight for Regional Development Network—A Practical Guide to Regional Foresight, Joint Research Centre of the European Commission, report EUR 20128 EN, December 2001.

"Fraunhofer Creates New Group for Innovation Research," Fraunhofer Society for the Advancement of Applied Research, July 3, 2017.

"Futures, Foresight and Horizon Scanning," Government Office for Science, https://foresightprojects.blog.gov.uk.

Gates, Bill, "A Better Response to the Next Pandemic," *GatesNotes*, January 18, 2010.

Gates, Bill, "The Next Outbreak? We're Not Ready," TED Talk, April 3, 2015.

"Global Market Share Held by Nokia Smartphones from 1st Quarter 2007 to 2nd Quarter 2013," Statista, July 25, 2013.

Global Trends 2025: A Transformed World, US National Intelligence Council, November 2008.

Global Trends 2030: Alternative Worlds, US National Intelligence Council, December 2012.

Global Trends: Paradox of Progress, US National Intelligence Council, January 2017.

Gorbis, Marina, "Five Principles for Thinking Like a Futurist," *Educause Review*, March 11, 2019.

Gorbis, Marina, *The Nature of the Future: Dispatches from the Socialstructed World*, 2013.

Grove, Andrew S., *Only the Paranoid Survive*, 1996.

Haines, Derek, "The E-Reader Device Is Dying a Rapid Death," Just Publishing Advice, January 9, 2018.

Handbook of Knowledge Society Foresight, European Foundation for the Improvement of Living and Working Conditions, 2003.

Harford, Tim, "Big Data: Are We Making a Big Mistake?," *Financial Times*, March 28, 2014.

Harris, Cassandra, *Small Countries, Big Innovations*, Strategic Business Insights, February 2018.

Hersh, Kenneth (moderator), "Closing Conversation with Jeff Bezos," 2018 Forum on Leadership at the George W. Bush Presidential Center, April 20, 2018.

Hinzmann, Brock, *Next-Generation Consumers*, SRI Consulting Business Intelligence, July 2009.

Hof, Robert D., "The Power of Us," *BusinessWeek*, June 19, 2005.

Hoffmann, Marcello, *eScience*, SRI Consulting Business Intelligence, November 2003.

Hollister, Sean, "The Age of the iPod Is Over," *Verge*, January 27, 2014.

"Horizon Scanning Programme: A New Approach for Policy Making," announcement by the UK Cabinet Office and Government Office for Science, July 12, 2013.

Kane, Gerald C., "Digital Disruption Is a People Problem," *MIT Sloan Management Review*, September 18, 2017.

Koppel, Ted, *Lights Out: A Cyberattack, A Nation Unprepared, Surviving the Aftermath*, 2015.

Kranzberg, Melvin, "Technology and History: 'Kranzberg's Laws,'" *Technology and Culture*, July 1986.

Krook, Joshua, "Why Experts Have Killed Innovation" (also published as "Innovation Is Dying, and Experts Are to Blame"), World Economic Forum, May 22, 2017.

Lindstrom, Martin, *Small Data: The Tiny Clues That Uncover Huge Trends*, 2016.

Mackintosh, James, "Three Economists Walk into a Bar," *Wall Street Journal*, January 9, 2017.

Madsbjerg, Christian, *Sensemaking: The Power of the Humanities in the Age of the Algorithm*, 2017.

Madsbjerg, Christian, and Mikkel B. Rasmussen, "The Power of 'Thick' Data," *Wall Street Journal*, March 21, 2014.

"Make No Mistake: To Err IS Human," *CBS News*, March 22, 2009.

Manyika, James, Michael Chui, Jacques Bughin, Richard Dobbs, Peter Bisson, and Alex Marrs, *Disruptive Technologies: Advances That Will Transform Life, Business, and the Global Economy*, McKinsey Global Institute, May 2013.

Marketing material, SRI International, circa 1980.

Martin, Roger, "Beyond the Numbers: Building Your Qualitative Intelligence," *Harvard Business Review*, May 1, 2010.

Maskell, Paul, *Music Consumer Insight Report 2016*, Ipsos, September 2016.

McGrath, Rita Gunther, "Is Your Company Ready to Operate as a Market?," *MIT Sloan Management Review*, June 29, 2016.

McGrath, Rita Gunther, *Seeing around Corners: How to Spot Inflection Points in Business Before They Happen*, 2019.

Mims, Christopher, "Five Tech Predictions for the Year Ahead," *Wall Street Journal*, December 28, 2015.

Morrison, Timothy, "Biodefense Summit Transcript," Public Health Emergency website, April 17, 2019.

Morson, Gary Saul, and Morton Schapiro, *Cents and Sensibility: What Economics Can Learn from the Humanities*, 2017.

Morvan, J. D., and Huang Jia Wei, *Infected*, Directorate-General for International Cooperation and Development (European Commission), January 31, 2012.

Novet, Jordan, "Amazon Says AWS Revenue Jumped 46 Percent in Third Quarter," CNBC, October 25, 2018.

O'Reilly, Tim, editorial review of *Race Against the Machine: How the Digital Revolution Is Accelerating Innovation, Driving Productivity, and Irreversibly Transforming Employment and the Economy*, Amazon.com.

"Oxford Dictionaries Word of the Year 2016 Is . . . ," Oxford Dictionaries, November 16, 2016.

Patton, Kermit M., *Reality Mining and the DataTurbine*, SRI Consulting Business Intelligence, June 2008.

Patton, Kermit M., "The Role of Scanning in Open Intelligence Systems," *Technological Forecasting & Social Change*, November 2005.

Paul, Fredric, "Traditional GPS Is Dead. Long Live Smartphone GPS," *Network World*, October 19, 2015.

Peng, Aster, *Community Values Trump Individualism*, SRI Consulting Business Intelligence, August 2009.

Peng, Aster, *Self-Tracking Health Data*, Strategic Business Insights, August 2011.

Pentland, Alex "Sandy," "Reality Mining," *MIT Technology Review*, March 2008.

Petrache, Ivona, *Halt the Epidemics!*, Strategic Business Insights, January 2017.

Pistrui, Joseph, "The Future of Human Work Is Imagination, Creativity, and Strategy," *Harvard Business Review*, January 18, 2018.

Ralston, Bill, and Ian Wilson, *The Scenario Planning Handbook: Developing Strategies in Uncertain Times*, 2006.

Ramírez, Rafael, Steve Churchhouse, Alejandra Palermo, and Jonas Hoffmann, "Using Scenario Planning to Reshape Strategy," *MIT Sloan Management Review*, Summer 2017.

Rohrbeck, René, *Corporate Foresight: Towards a Maturity Model for the Future Orientation of a Firm*, 2010.

Rosling, Hans, *Factfulness: Ten Reasons We're Wrong about the World—and Why Things Are Better Than You Think*, 2018.

Saffo, Paul, "Six Rules for ~~Accurate~~ Effective Forecasting," *Harvard Business Review*, July–August 2007.

Safian, Robert, "This Is Generation Flux: Meet the Pioneers of the New [and Chaotic] Frontier of Business," *Fast Company*, January 9, 2012.

Samuels, Michelle, "Depression Rate Has Tripled among US Adults," Brink, Boston University, September 2, 2020.

Sargut, Gökçe, and Rita Gunther McGrath, "Learning to Live with Complexity," *Harvard Business Review*, September 2011.

Schoemaker, Paul, with Robert E. Gunther, *Profiting from Uncertainty: Strategies for Succeeding No Matter What the Future Brings*, 2002.

Schwartz, Peter, *The Art of the Long View: Planning for the Future in an Uncertain World*, 1996.

Schwirn, Martin, *Automation on Top of the Value Chain*, Strategic Business Insights, June 2019.

Schwirn, Martin, *Autonomous Robots in the Wild*, Strategic Business Insights, April 2012.

Schwirn, Martin, *Diagnosed Self*, Strategic Business Insights, November 2014.

Schwirn, Martin, *A Different Kind of "Internet of Things,"* Strategic Business Insights, February 2011.

Schwirn, Martin, *From Cocreation to Competition*, SRI Consulting Business Intelligence, September 2005.

Schwirn, Martin, *Leveraging Photography on the Web*, SRI Consulting Business Intelligence, January 2008.

Schwirn, Martin, *Life after the Time of Coronavirus*, Strategic Business Insights, April 2020.

Schwirn, Martin, *Quantified Employees*, Strategic Business Insights, January 2016.

Schwirn, Martin, *Quantified Everybody*, Strategic Business Insights, June 2014.

Schwirn, Martin, *Unexpected Competition from Smartphones*, SRI Consulting Business Intelligence, March 2009.

Schwirn, Martin, *URLs for Products: The Internet of Things*, SRI Consulting Business Intelligence, June 2003.

Silver, Nate, *The Signal and the Noise: Why So Many Predictions Fail—but Some Don't*, 2012.

Smith, Fred, and Brian Dumaine, "How I Delivered the Goods," *CNN Money*, October 1, 2002.

Smith, James B., and Pamela G. Kruzic, *Analyzing Future Business Environments*, SRI International's Business Intelligence Program, 1976.

Swift, Taylor, "To Apple, Love Taylor," Taylor Swift Tumblr account, June 2015.

Taleb, Nassim N., "Beware the Big Errors of 'Big Data,'" *Wired*, February 8, 2013.

Telford, Carl, *Quantified Communities*, Strategic Business Insights, July 2014.

Telford, Carl, *Robot Representation*, Strategic Business Insights, June 2011.

Tetlock, Philip, and Dan Gardner, *Superforecasting: The Art and Science of Prediction*, 2015.

Thompson, Helen, "The Coming Crisis: We're Not in Kansas Any More," Sheffield Political Economy Research Institute, May 2016.

Tighe, Steve, *Rethinking Strategy: How to Anticipate the Future, Slow Down Change, and Improve Decision Making*, 2019.

Toffler, Alvin, *The Third Wave*, 1980.

Watson, R. T., "Trading Places: Global Box Office Dethroned by Spending on Home Entertainment," *Wall Street Journal*, March 21, 2019.

Webb, Amy, "A National Office for Strategic Foresight Anchored in Critical Science and Technologies," Working Paper, October 17, 2019, Freeman Spogli Institute for International Studies, Stanford University.

Webb, Amy, *The Signals Are Talking: Why Today's Fringe Is Tomorrow's Mainstream*, 2016.

Weick, Karl E., and Kathleen M. Sutcliffe, *Managing the Unexpected: Sustained Performance in a Complex World*, third edition, 2015.

Wiesbrock, Kimberly H., *Collaborative Consumption*, Strategic Business Insights, December 2010.

Wijman, Tom, "Mobile Revenues Account for More than 50% of the Global Games Market as It Reaches $137.9 Billion in 2018," Newzoo, April 30, 2018.

Wikipedia, online entry, Wikimedia Foundation, June 7, 2018.

Wilkinson, Angela, and Roland Kupers, "Living in the Futures," *Harvard Business Review*, May 2013.

World's Greatest Brands, Interbrand, 1996.

Zambon, Kat, *Global Health Experts Advise Advance Planning for Inevitable Pandemic*, Georgetown University Medical Center, January 12, 2017.

Zetlin, Minda, "Blockbuster Could Have Bought Netflix for $50 Million, but the CEO Thought It Was a Joke," *Inc.*, no date.

Zhang, Michael, "This Latest Camera Sales Chart Shows the Compact Camera near Death," PetaPixel, March 3, 2017.

Quotations

Ali, Muhammad, American boxer: "Float like a butterfly, sting like a bee. The hands can't hit what the eyes can't see."

Amara, Roy Charles, American researcher: "We tend to overestimate the effect of a technology in the short run and underestimate the effect in the long run."

Bohr, Niels, Danish physicist: "Prediction is very difficult, especially about the future."

Einstein, Albert, German physicist: "Imagination is more important than knowledge. For knowledge is limited."

Fitzgerald, Scott F., American writer: "The test of a first-rate intelligence is the ability to hold two opposed ideas in the mind at the same time, and still retain the ability to function."

Gerstner, Lou Jr., former CEO of IBM: "Culture isn't just one aspect of the game; it is the game."

Gibson, William, American-Canadian writer: "The future is already here—it's just not very evenly distributed."

Gibson, William, American-Canadian writer: "The street finds its own uses for things."

Gretzky, Wayne, Canadian ice hockey player: "A good hockey player plays where the puck is. A great hockey player plays where the puck is going to be."

Heraclitus of Ephesus, Greek philosopher: "If you do not expect the unexpected, you will not find it."

Olsen, Ken, founder of Digital Equipment Corporation: "There is no reason anyone would want a computer in their home."

Osler, William, Canadian physician: "The value of experience is not in seeing much, but in seeing wisely."

Pasteur, Louis, French microbiologist: "In the fields of observation chance favors only the prepared mind."

Toffler, Alvin, American futurist and author: "Our obsessive emphasis on quantified detail without context, on progressively finer and finer measurement of smaller and smaller problems, leaves us knowing more and more about less and less."

Tukey, John, American mathematician: "The greatest value of a picture is when it forces us to notice what we never expected to see."

Tyson, Mike, American boxer: "Everyone has a plan until they get punched in the mouth."

Unknown (often attributed to Albert Einstein, German physicist, or Elliot Eisner, American professor at Stanford University): "Not everything that matters can be measured, and not everything that can be measured matters."

Watson, Thomas, former president of IBM: "I think there is a world market for maybe five computers."

Wittgenstein, Ludwig, Austrian-British philosopher: "If people did not do silly things, nothing intelligent would ever get done."

Wolfe, Jeremy, professor at Harvard Medical School and principal investigator at the Visual Attention Lab: "If you don't see it often, you often don't see it."

Yoda, fictional Jedi Master in motion picture *The Empire Strikes Back*: "Difficult to see. Always in motion is the future."

Index

L

Lab126, 66
Lansdale, Edward, 147
leaders, training of future, 172
Lederman, Mara, 105
Letter to Shareholders (Amazon), 106
Leveraging Photography on the Web
    (Schwirn), 45, 90
LGBTQA community, 164
Life after the Time of Coronavirus
    (Schwirn), 120, 122
light-emitting diodes (LEDs), 48
*Lights Out: A Cyberattack, A Nation
    Unprepared, Surviving the Aftermath*
    (Koppel), 64
Li Keqiang, 68
Lindstrom, Martin, 68
long tail, concept of, 61
"The Long Tail" (Anderson), 61, 63
Loungani, Prakash, 18
Lyft, 97

M

machine learning, 78, 92, 108, 146
Madsbjerg, Christian, 9–10, 105, 106
magazines, as information source, 63
maker movement, 163–164
Manhattan Project, 117
Marina, Gorbis, 29, 81
market capitalization, 97, 127
market development, 166
marketing material, SRI International's, 16
market research, 2, 35, 151–152, 173
market share, 108, 113
Martin, Roger, 10, 38–39, 149
Massachusetts Institute of Technology, 64,
    70, 75, 83, 92, 96
McAfee, Andy, 70, 75, 76, 78
McDonnell, Edward, 81, 128–129, 136,
    151
McGrath, Rita Gunther, 17–18, 37, 74,
    146
McKinsey & Company, 79, 143
McKinsey Global Institute, 70, 76, 83
McNamara, Robert, 147

meaning, identifying, 108–126
    by exploring new market environ-
        ments, 123–124
    impact/emergence matrix for, 113–115
    meaningful information, 72, 73
    myopic view for, 125
    in a nutshell, 126
    by prioritizing issues, 111–113
    role of scanning in, 108–109
    to take action, 129–132
    technology and, 109–111
    wild cards and, 116–123
Mechanical Turk, 32
Media Lab, MIT, 96
medical research, 174
MERS coronavirus, 119
Me Too movement, 164
Microsoft, 3, 57, 65, 92, 119–120, 156
Microsoft Encarta, 57
Middle East respiratory syndrome
    (MERS), 119
milestones, 161
Mims, Christopher, 83
*Mitigating the Impact of Pandemic
    Influenza through Vaccine Innovation*
    (US Council of Economic Advisers),
    120
"Mobile Revenues Account for More than
    50% of the Global Games Market"
    (Wijman), 100
Moore, Gordon, 23, 66
Moore's law, 23, 66
Morrison, Timothy, 120
Morson, Gary Saul, 10, 18, 88, 104
Mosaic, 33
"Mother of All Demos" (Engelbart), 65
Motorola, 3, 4, 156
MP3 music files/players, 33, 54–55, 100
Ms. Swift's love letter, 66
Musk, Elon, 66
Muvee, 70, 76, 92

N

nanotechnology, 20, 157
Napster, 33, 35, 159
narrativeness, 104–105

World Trade Center terrorist attacks, 19–20, 117
World Wide Web, 27, 30, 33, 65, 174

X

X (company), 143
Xerox, 65, 159

Y

Yahoo!, 140
Yoda (fictional Jedi Master), 179
YouTube, 35, 50, 110

Z

Zhang, Michael, 100

About the Author

Martin Schwirn is a vice president at Strategic Business Insights, an SRI International spinoff. He is focused on strategic and innovation-related consulting, including foresighting, horizon scanning, and scenario planning. He is the director of the scanning methodology that *Small Data, Big Disruptions* expands upon and has worked internationally with the process for more than two decades. He has helped companies from virtually every industry, a wide range of organizations, and many government departments in Asia, Europe, and North and South America to anticipate disruptions and change. He lives in San Francisco, California, and works in Menlo Park, Silicon Valley.

Schwirn has dedicated the past two decades of his professional life to horizon scanning to find the crucial changes in today's marketplace that will shape tomorrow's world. He has led and participated in hundreds of workshops to find strategically crucial developments and to make sense of them. In dozens of scenario-planning and roadmapping projects he has helped strategists to immerse themselves in future opportunities and challenges. He therefore understands how horizon scanning fits in the tool set of future-oriented initiatives. On four continents and in more than a dozen countries, he has experienced the differing expectations that participants from organizations in different industries have and how they think about emerging trends and disruptive developments. He has seen how businesses, organizations, and government-related entities think about the future and what priorities they set when looking toward tomorrow's challenges.